$\mathcal{A}$
SPARKLE
in
Their Eyes

A
SPARKLE
in
Their Eyes

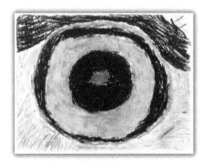

Raising Talented, Diverse Students
in STEAM Careers

Dr. Veronica A. Wilkerson-Johnson

MILL CITY PRESS

Mill City Press, Inc.
2301 Lucien Way #415
Maitland, FL 32751
407.339.4217
www.millcitypress.net

Paperback ISBN-13: 978-1-66287-149-8
Ebook ISBN-13: 978-1-66287-150-4

Table of Contents

Dedication

THIS WORK IS dedicated to my parents, James Henry Wilkerson and Alberta Dixon Wilkerson, Registered Nurse (RN). My mother was our family's first-generation student, and the first STEM professional in my life. She taught me, lovingly and by example, that one can succeed against all odds as an underrepresented student or, as I will also refer to it, a diverse student, or student of color. She reminded me, as she often said and sometimes sang, "If I can help somebody as I travel along, then my living shall not be in vain." This primer is also dedicated to my wonderful family who each encompass roles in the worlds of STEM/STEAM Arts (science, technology, engineering, the arts, and math). My husband Melvin, (a military veteran, worked in Public Safety and Environmental water systems) who inspires our family and me every step of the way. Our amazing children, STEM heroes all: Dichondra, outstanding entrepreneur, artist, realtor, and business and public policy strategist (thank you for sharing the Hamtramck Parks Conservancy STEAM Camp experience); Christopher, talented architect and Airport Planner at Passero Associates, LLC (thank you for your Conceptual Airport Design), daughter-in-law Akinya Joy Moore-Johnson, talented Dance and Yoga Instructor, and Business/Marketing Executive, and son Jonell Henry Johnson, Health Care professional and community advocate, and our family's Poet Laureate (thank you for the poetic tribute to your father!).

Furthermore, our grandson Miles Michael Geiger, future STEAM professional, whose 4th grade art design, adorns the cover of this primer. Thank you, Miles! Our nephew, Fredrick Johnson, the Design Artist who created the beautiful drawings of *Parents We Love*, and *Chicken Charlie and His Brothers,* you have made them more dear! To my grandmother, Leonia Bell Dixon, and ancestors who taught me much about natural healing and the power of plants and nature. To our beloved families–Pinkston, Nash, Dopson, Ellis, Green, Broomfield, Bradley, Elliott, Jones, Geiger, Perkins, Birton, Favor, Moore, Bramwell, Blake, Estes, English, Barbin, and Duckworth, Love you all!

My Team Johnson is my constant ROCK and source of support, love, and inspiration. I love you all! My dear ones of the City of Inkster, the "Village" that was my incubator. In that village I met lifelong friends and classmates—thank you Dr. Deborah Smith Card–teachers who inspired and helped shape my life, Mrs. Gloria C. Ratliff, my 5th-6th grade teacher mentioned later in this primer, High School teachers, Mr. Otis Simmons, English, Mrs. Maude Reid, French, Mr. Edwin Ames, Music (I played viola in the orchestra), and Mr. Fitzhugh Houston, Math. As well as Mr. William Daniels, Chemistry and Driver's Education teacher all in one, and who later became Inkster's Mayor.

The many dear people of Womack Temple C.M.E. Church who inspired me in all things education and spiritual fulfillment. There I was inspired by teachers as Mrs. Rosa Nash, and mentored in piano and viola by the music director, Mr. Andrew Lester, Madeline Lester and their daughter, my mentor, Wilbie Lester Cobb, and her husband

Willie Cobb, fellow members of the Churches of Christ. Thank you to my friends, Author and Minister Karen Williams; Dr. Lillian Brooks; Sandra Kay Cehr, Environmental and Health Sciences Advocate; and Jahton H. Bishop, Graphic Artist, A Cappella Gospel Express Radio Host, and Founder/Director of the Jesse H. Bishop Crusaders A Cappella Choir.

I could go on extensively naming the wonderful people who helped shape my career and life, as now I seek to assist the next generations in uplifting theirs, for it is time for all diverse people to fulfill their dreams in STEM and the Arts—and embrace the Sparkle in Their Eyes.

I hope this narrative helps you as parents think of ways you are building legacies with and for your children, as you remember those who lovingly helped you build yours. Parents, family members, and mentors of diverse and first-generation students, it is a great adventure being a part of the Sparkle in your Children's Eyes. Certainly, in the wave of change, we will find our direction.

I also thank those who inspired my educational journey, Dr. Roberta Teahen and Dr. Sandra Balkema who encouraged my doctoral achievement at Ferris State University. I am forever grateful to you and the Ferris DCCL faculty. Additionally, to my amazing Dissertation Committee, it was such a blessing and an honor to work with you, and I thank you for your astute leadership, expertise, and encouragement in mentoring me through my doctoral journey. Dr. Jack Becherer, my wonderful Dissertation Committee Chair, Dr. Daniel Burcham, your guidance and candor meant so much, and Dr. Kenneth Slater, humanitarian leader whose friendship continues in my professional, church and spiritual walk. Thank you for all that you do, and for believing in me.

Furthermore, I want to thank my University of Michigan mentor, Ms. Josivet Moss, and the awesome faculty and staff of the University of Michigan School of Information (UMSI). UM Professor Thomas Slavens said to our class as we completed our Master's Program at UM; remember the children's story of Goldilocks and the Three Bears. He said, Papa Bear acknowledged that there was an invader related concern—"Someone's been sitting in my chair," he said. Baby Bear agreed and analyzed the outcome—"Someone's been sitting in my chair too, and they broke it all down!" However, in all three instances regarding their porridge, chairs and beds being invaded by Goldilocks, Mama Bear simply went on repeating 'mine too' in each instance. While she reacted, as did the others, she did not discover, add to, or offer any further insight at any point during the discomforting discussion. Professor Slavens concluded—whatever you do, in your careers and in life, do not become Mama Bear professionals. Contribute proactively and meaningfully of your talents to the world, and make a difference wherever you go. I will always be thankful for the UM School of Information, its great faculty and staff. I am also proud of the UMSI endowment that bears my name, the Dr. Veronica Wilkerson Johnson Diversity Internship Award.

Thank you to all of my many dear friends and distinguished leaders and colleagues in Michigan state government, Wayne County, the Library of Michigan, Lansing Community College, and the UM Office of the Vice President for Government Relations, Ann Arbor, Michigan! Your teamwork makes the dream work!

Thank you to Michigan State University, the wonderful faculty and fellow graduates of my Bachelor of Arts (BA) Program at Wayne State University College of Fine, Performing and Communication Arts, and the Wayne State University College of Nursing that honors my mother's legacy through the Alberta Dixon Wilkerson, RN Endowed Diversity Scholarship Award.

***

"Change will not come if we wait for some other person or some other time. We are the ones we've been waiting for. We are the change that we seek."

President Barack Obama

Introduction

Sometimes in the wave of change, we find our direction.

A Primer for Parents of Underrepresented Students Seeking STEM/STEAM (Science, Technology, Engineering, the Arts, and Mathematics) Careers

IT IS MY pleasure to invite all parents of diverse, first-generation, and female students seeking STEM/STEAM (Science, Technology, Engineering, the Arts, and Math) careers, to explore the sheer joy of learning as you inspire excellence in your children. The delight is that learning is a lifelong gift we all have been given. How we use it is up to us. Remember that you are your children's Hero! They look up to you, follow your guidance, and they inevitably will "live, refresh, repeat, and recite" what you share with them. Teach them well. Our children are our future, and remember, as Picasso said, "every child is an artist."

This book is a culmination of my doctoral research and my quest to fill the world with diverse students of excellence in STEM and in the Arts.

Therefore, this journey begins, and I preface this work with a message about my mission in exploring the importance of diversity in the many fields in STEM and STEAM. Science, Technology, Engineering, the Arts and Math, have existed since time began. We now talk more about them, for they are the basis of much that we enjoy in modern society. Yet, there is inequity in the numbers of people of color in STEM/STEAM professions. African American, Native American, Hispanic, Asian, Arabic and Pacific Islander populations, and women of all cultures, still lag behind in taking their rightful places in the innovative fields of STEM and the Arts.

The personal narratives I share in this primer are a way to inspire you as parents to build meaningful experiences, and gather people and organizations into your network that will inspire your children to academic excellence and career success. Teach them to act as though there is no reason to be afraid. Surround them with teachers, resources, and support that build their confidence to stretch out on their inherent talents and dreams, for there is genius in every one of us.

Many students aspire to STEM/STEAM careers, and we know that, historically, STEM and the Arts careers have revolutionized the world, and that STEM professionals make many of the current advancements at the forefront possible. As we experience more change in our world, climate change, health care, and international concerns, the United States is hoping that STEM professionals will aid the many social, economic, technological, and national security needs we will face. Every ethnic group in the U.S. should have a proportional representation in the STEM workforce.

The Changing American Family

By census numbers, America is far more diverse now than in the past. From the 2020 census, the U.S. Census Bureau reports that 4 out of 10 Americans identify as a race or ethnic group other than white. The number of females in the U.S. as of July 2019 was 166.6 million, compared to 161.7 million men. As noted by William H. Frey in his article "The nation is diversifying faster than predicted, according to new census data," Brookings, 7-1-20, since 2010, racial minorities accounted for all U.S. population growth, especially with the increase in the Latino and Hispanic population. In 2019, for the first time in American history, more than half of the nation's population under age 16 identified as a racial or ethnic minority. The 2020 Census reflects that the U.S. grew by 19.5 million people between 2010 and 2019, a growth rate of 6.3%. While the white population declined a fraction of a percent, Latino and Hispanic, Asian American, African American and Native American populations grew by 20%, 29%, 8.5%, and 7.6%, respectively. Moreover, those identifying as being two or more races grew by 30%. While immigration accounted for 74% of Asian growth, and 24% of Latino or Hispanic growth, the population increase among people of color has primarily occurred by natural births. Now, Latino or Hispanic and African Americans together comprise almost 40% of the U.S. population.

Although women outnumber men in every age group, they do not earn as much as men, and do not comprise a significant percentage of the STEM/STEAM workforce. According to the U.S. Census Bureau, as of 2019, 79.2 million women, ages 16 and older, were working full-time, year-round jobs, earning 81.6% of what men made. As of 2022, more women are graduating in STEM careers than in previous decades. The gender distribution in STEM fields varies by subject, according to Forbes, 4-7-22. Compared to men, women represent 16% of Bachelor's degrees in computer and information sciences, 21% in engineering and engineering technology, 27% in economics, and 38% of recipients in physical sciences. On the other hand, women represent 48% of Bachelor's degrees in mathematics and statistics,

63% in biological and biomedical sciences, and 83% in health professions and social and behavioral sciences. Additionally, according to the Center for First-Generation Student Success, National Data Fact Sheet, 56% of college undergraduates in the U.S. are first-generation students, and 59% of these are also the first sibling in their family to go to college. Of these first-gen students, more each year are pursuing STEM education.

Although the numbers of Hispanics, African Americans, and Native Americans have increased in the population as of 2022, research shows that Hispanic workers comprise only 8% of the national STEM workforce, up by only 1% since 2016, and African Americans are only 9% of STEM professionals in the U.S., as reported by the PEW Research Center.

Inevitably, these changes in the proportional numbers of diverse populations in America, and comparative diversity and gender representation in the STEM workforce, represent opportunities for people of color, first-generation and women to pursue their dreams in STEM/STEAM education and careers.

Much is also changing in the world of community colleges, universities, and in technical training programs, bringing tremendous opportunities for your children. Now is the time for you to encourage and help them embark upon this remarkable journey.

Studies show that parental support, encouragement, intervention, and guidance greatly help in creating stimulating, language-rich, supportive environments that defy the odds of socioeconomic circumstances. Loving family relationships are the key. Family patterns of life, relationships, practices, and concerns need not hinder your efforts as parents and mentors to improve the curriculum of the home. Times well spent with your children in reading, studying, exploring and learning together will do wonders for your children's academic progress. There are many programs, resources, and types of assistance available to parents of students of color, women, and first-generation students. In this primer and beyond are comprehensive resources to aid your students in their quest for excellence in STEM and STEAM careers.

Background

As we have noted, we need more and better STEM and Arts education available to all. While exploring this concern in my doctoral research, I reviewed the literature and decided that it is time to empower our families and communities to mentor tomorrow's STEM giants. Together, let us create a better opportunity for our student's academic and career success!

This primer will provide ideas, programs and resources on the STEM/STEAM opportunities, where to get further information, how to communicate with the faculty and administrators of educational institutions, including early childhood development centers, K-12 schools, colleges, and STEM/STEAM technology training centers. It will offer ways that you as parents can support and encourage your children in the pursuit of STEM/Arts education and activities, and academic and cultural resources, regardless of where you live, that are available through your student's instructors, counselors, gifted and talented programs, community resources, local community colleges and universities, and parent groups. A wealth of programs, communication approaches, and enrichment activities can aid in the development of your children's STEM/STEAM interests.

This primer will provide:

- Background on the underrepresentation of people of color, first-generation students, and women in STEM/Arts education and professions.

- Descriptions of the problems that the parents of underrepresented students face when attempting to be advocates for their children, particularly when the parents feel their own educational concerns.

- Recommendations for interpersonal tools that will help STEM/STEAM parents feel comfortable in the educational settings that their children will encounter, helping them to understand the differences in social communication, "class," and "social register" language school personnel may give.

- A wealth of sample school programs that offer underrepresented parents the types of preparation their children will need to succeed in STEM/Arts education and professions.

- Role-play exercises and trigger questions that encourage parents to become more aware of their student's school and community settings, and to mentor them in communication skills and strategies.

- Information regarding the STEM/STEAM education pipeline that is awaiting their children that help mentor and aid students of color, first-generation students, and women.

- Emphasis on the importance of school personnel, counselors, mentors, and professional role models in your children's educational development is clear. You can engage diverse researchers, engineers, and educators in STEM fields to coach, mentor, and tutor your children. As more parents become empowered to develop their children in STEM education, the emergence of a new cohort of STEM-trained professionals will result, and this will help fill the STEM/STEAM pipeline and transform STEM/STEAM communities.

- Encouragement for you to consider the benefits of pre-K and pre-college training in math and science so that your children are prepared for STEM/STEAM educational opportunities.

- Emphasis on the importance of retention, as this primer encourages you and all STEM parents to keep your children focused and attentive to their STEM/STEAM education and career goals. Research confirms that there is a high attrition rate among students of color who choose to drop out when their STEM courses become difficult. Some give up, drop out, or change their majors. Help your students remain steadfast and firmly committed to staying the course and ultimately completing their STEM/STEAM education goals!

"Strong, confident, compassionate parents raise strong, confident, compassionate kids."

Early Teachings

When I was two years old my mother began immersing me in books, some colorful early readers, some early math, and a particular lovely, old leather-bound set that I have in my personal library to this day, entitled "Childcraft in Fourteen Volumes", published by Field Enterprises, Inc., Chicago, IL. I could not have known at the age of two, how much these books would inspire my imagination and understanding of the world and education. The set included children's short stories, poems, early learning, and stories from around the world, and it included volumes dedicated to assisting parents and teachers as they help students develop everything from the care of a newborn baby, cognitive and healthy development of toddlers and preschoolers, to guiding pre-adolescent children up to 13 years of age. I can remember looking through the books, enjoying my parents reading me the stories, and sharing the beautiful imagery on those pages. I realize now that there were few pictures of children of color in those books, but to my parents' credit, that did not hinder them from exposing me to the joy of reading, learning, imagining, and exploring with these books and many others.

"Carpe diem—Seize the day!"

Horace, Roman poet

Interesting Takes on Life

In addition to the wonderful stories for children in these books, parents could learn what to expect in school settings and from teachers. What emotions speed up or aid learning, and how happy families doing things together aid children's emotional development. In addition to, exploring symbol and number association, "doing" hobbies, and school and community activities to advance social development, and training on all things parenting, socialization, and learning. It is historically clear, that expanding our knowledge as parents will also inspire the minds of our children. It is joyful and fulfilling for you and your children to learn and experience together.

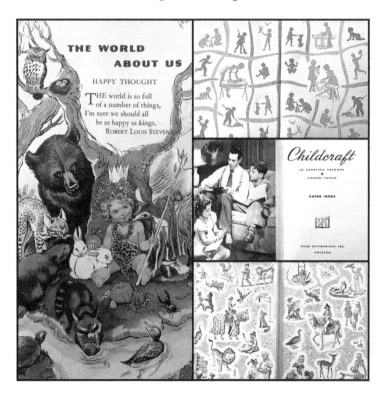

Photo images, "Childcraft in Fourteen Volumes," 2022

A Tribute to Mrs. Gloria C. Ratliff

The Teachers Who Shape Us Are the
Kindness That Makes Us

"Joy is not in things, it is in us."

Charles Wagner

In this primer, I mention a number of outstanding educators who helped to shape my academic and career pathways. One to whom I give special praise is Mrs. Gloria C. Ratliff, my 5th and 6th grade teacher at the former Woodson Elementary School in Inkster, Michigan! A lot of our grade school success was due to Mrs. Ratliff, who taught us we could be anything we wanted to be. We could dream big, and as she said, "learn well and become excellent in all that you do." Mrs. Ratliff taught all areas of the curriculum, and she was one of our first STEM/STEAM instructors.

Coming a distance each day from West Bloomfield, Michigan, Mrs. Ratliff brought an aura of sophistication and she taught us about innovations in affluent Oakland County, MI that we had not experienced in our small town of Inkster, a 6.25 sq. mi. city, 17 miles west of Detroit, Michigan, and 26 miles from West Bloomfield.

To provide a brief history of Inkster, according to the 2020 U.S. Census, Inkster, Michigan has 26,088 residents: 73.2% African American, 20.5% White, 2.6% Hispanic or Latino, 1.6% Asian, 0.3% Native American, 0.1% Pacific Islander, and 1.7 other. As noted by Alice J. Bostick, author of "The Roots of Inkster", c.1980, it was historically called Moulin Rouge (or 'Red Mill") by Native Americans and early settlers in 1825. The Moulin Rouge Post Office was established in 1857, and in 1863, a Scotsman, Robert Inkster, (Norse name Ingasetter), born March 27, 1828 in Lerwick, Shetland, arrived and began operating a steam sawmill on present day Inkster Road near Michigan Avenue. The post office was renamed Inkster in July 1863, and the Village of Inkster was incorporated in 1926. The present City of Inkster was incorporated in 1964.

It is important to note that, the Inkster Public Schools were dissolved in July 2013, due to financial concerns. Therefore, Inkster K-12 students now attend schools in geographic proximity to their Inkster locations – going to Wayne-Westland Community Schools, Westwood Community Schools, Taylor School District or Romulus School District, respectively.

Again, historically, before the abolition of slavery, Moulin Rouge/ Inkster was a stop on the Underground Railroad, at Wesleyan Methodist Church which once stood at the corner of Michigan Avenue and Henry Ruff Roads. Many streets in Inkster are named after early farmers who settled in Moulin Rouge/Inkster and cultivated farms in the 1800's to the early 1900's, including the Ruff, Daly, Harrison, and Inkster families. In the 1920's–1930's, the boom of the industrial era, and the Henry Ford automotive plant development in Dearborn, Michigan, attracted many African Americans and diverse cultures to the Inkster area from the South and other states. Henry Ford paid competitive wages at that time, $5/day for a 40-hour work week, and he helped to establish affordable housing communities in nearby Inkster for the workers. Endearing names emerged over the years as thousands of people came to work in the automotive industry, and populated the housing projects in Inkster. "Cardboard City", Carver Homes, and LeMoyne Gardens were a few. Research shows that by 1930 a car existed for every 5 Americans, and Inkster became a hub of heavy traffic along U.S. 12 (Michigan Avenue), as many traveled to and from Detroit and the Dearborn auto plants.

A city of proud residents, some of the many notables of Inkster are:

* Geraldine Hoff Doyle, often credited as being the model for the "Rosie the Riveter" World War II era "We Can Do It" poster, promoting female war workers aiding industry and morale, was born in Inkster on July 31, 1924,

* U.S. Congressman Vern Buchanan (R, 16th District, Florida), graduated from Inkster High School in 1969,

* Jewell Jones, youngest person to be elected to the Inkster City Council, and the Michigan House of Representatives, serving as State Representative (D-District 11) for three House terms, 2017-2022,

* Jeralean Talley, (May 23, 1899-June 17, 2015), was the world's oldest living person until her passing at 116,

* Earl Jones, U.S. Olympics Bronze Medalist, middle distance runner in the 1984 Olympics,

* Tyrone Wheatley, NFL football offensive tackle and coach, MVP of 1993 Rose Bowl, University of Michigan Wolverines,

* Keshawn Martin, NFL football wide receiver, college football, Michigan State University Spartans,

* Marcus Fizer, former NBA basketball player,

* J'Leon Love, American professional boxer, Super Middleweight division,

* Wade Flemons, R&B singer, wrote "Stand by Me" sung by The Platters, and was a member of Earth, Wind and Fire until 1973,

* The original Motown Marvelettes – Gladys C. Horton, Wanda Young, Juanita Cowart, Katherine Anderson, Georgeanna Tillman, and Georgia Dobbins.

Having reflected a bit on the background and history of Inkster, Woodson Elementary School was located between several of the housing projects where most of us resided, notably Cardboard City and LeMoyne Gardens. We looked forward to school each day and to showing our teacher, Mrs. Ratliff, our latest homemade creations. As children, we did not think about, or understand, the complexities of "class" comparisons. Where others might have assumed our small, blue-collar city, and schools, were inferior to neighboring, more privileged communities, somehow, socioeconomic status was not our rite of passage, nor our interest. We were happy and contented being nurtured and loved, knowing that we had adventures to bring us joy, and friends to share them with.

When we arrived each day to Mrs. Ratliff's classroom, she bolstered our sense of value and self-esteem. She asked about the fun things we enjoyed in our rural adventures. The tadpoles, newts and salamanders we found around ponds that became our pets. The "Butter Candy" we made at home that our mothers let us bring to school, carefully wrapped in wax paper and paper bags. The jacks and ball, marbles, and other games we competitively played. The hu-la-hoops and jump rope "Double Dutch" games. The acorns we harvested to make necklaces. And the penny candy we saved up to buy and enjoy from Logan's Store, Hearn's Market, or Allen's Supermarket. Mrs. Ratliff cared about our educational development, and about our emotional well-being. Beyond assuring that we could do the basic "read, write and arithmetic". she brought an unconditional and nurturing devotion to teaching that made learning exciting and fun. She always had special ways of inspiring our curiosity on all things academic.

One day, Mrs. Ratliff brought us beautifully packaged tins from her family trip to Hawaii. She enthralled us with Hawaiian stories and descriptions of natural habitats in tropical, Polynesian islands. As she did so, she opened the tins to reveal fragrant macadamia nuts, and each of us bright-eyed children got to taste and enjoy the delicious, lush treats. She asked what we thought about them, how we liked them, knowing that most of us had never tasted macadamia nuts before. What are some adjectives to describe them, she asked? Buttery, succulent, smooth, salty? Where and how do they grow to develop such velvety texture? How do you roast them? We savored the nuts for a while, immersed in her stories, while images of Hawaii and beautiful islands danced in our heads.

From that moment to this day, I love macadamia nuts. The history, natural wonders and culture of Pacific islands all enthrall me, because of Mrs. Ratliff. From her I learned that teachers play such a vital role in the sights, sounds, and images that ignite our imagination, and this is especially true of STEM and the Arts teachers and mentors.

Mrs. Ratliff made me valedictorian of our 6th grade graduating class at Woodson, and she encouraged me when I asked to write a graduation song. I completed the song for our class to sing, the chorus concluding, "We're going to leave you Woodson School, going on to Fellrath…" Our 6th grade graduation was beautiful, and Mrs. Ratliff's inspirational teaching and uplifting voice sent us on our way, ready to reach new heights at Inkster's Fellrath Junior High School. Clearly, many of us have proceeded well upon our journeys and paths in life since then, and in our careers, because of her.

I never forgot Mrs. Ratliff's amazing outlook and the ideas she instilled. Years later, I reached out to find and thank her. Mrs. Ratliff's wonderful family–husband, the late West Bloomfield Judge Carl T. Ratliff, Sr., and her outstanding children, Risa Lynn Ratliff, MA, Psychology, who is a Life and Motivation Coach in Farmington Hills, Michigan, and Dr. Carl T. Ratliff, Jr., a Psychiatrist in Kokomo, Indiana (both STEM professionals!) welcomed my thankful visit, and shared in my joy, as my teacher and I sat, and spent a wonderful day reminiscing, and celebrating all that her great inspiration had helped me to achieve. I will always appreciate Mrs. Ratliff's contribution to teaching, and her awesome children and family, who carry on her legacy. I also thank them for permitting me to highlight her in this story, and honor her in this heartfelt tribute.

I hope this will inspire teachers everywhere, especially on those days when you wonder if going that extra mile for students is making a difference. Believe me, it is. I also hope it will inspire parents, as you seek educators who will propel your children to academic excellence. As well as, I hope that every student reading this will be blessed to have teachers and mentors like Mrs. Gloria C. Ratliff, who will inspire you along your STEM/STEAM journey, and one day help you make an exceptional difference in the career you pursue.

If you are a STEM/STEAM educator, a diverse parent, or a mentor, remember, the students you are guiding today will be our leaders and STEM/STEAM professionals of tomorrow. Teach them well.

Mrs. Gloria C. Ratliff, 5th and 6th Grade Teacher at
Woodson Elementary School, Inkster, Michigan
Photo Courtesy of Ms. Risa Lynn Ratliff, West Bloomfield, Michigan, 2021

***

The Thought, the Method, the Transformation

What does it mean to inspire in STEM, Entrepreneurship, and the Arts?

"Ah, but a man's reach should exceed his grasp. Or what is a heaven for?"

Robert Browning

THIS PRIMER IS designed to help ignite your imagination, as diverse families, about the many new advancements happening all around us that your children will love to explore. Your STEM stars may one day help in the development of transportation marvels that today would defy our wildest imaginations. Here are examples of University of Michigan and Michigan State University driverless or autonomous vehicles that will be transporting many. Take your students for a ride in these and many other autonomous vehicles quickly emerging around the globe. Moreover, other stories to follow will share how our amazing world is quickly becoming the "future" that the animated Jetsons series portrayed, back in 1962. Whether it is planning the airports of the future, the ways that we understand outer space and the depths of our oceans, or how we explore other parts of our planet or universe, you can help your children become a contributing part of this innovative next generation.

"There's a way to do it better, find it."

Thomas Edison

Let Us Prepare Our Students for Next Generation Innovations

As we consider the many careers in aviation, aeronautics and beyond, we are aware of how our world is evolving technologically. Driverless or autonomous aircraft are being designed right alongside driverless cars and other modes of transportation. Your students may ask who creates the infrastructure that supports these innovations. Just as roads and highways are changing to accommodate increased traffic and types of vehicles, so airports must evolve.

Following is a conceptual AutoCAD (computer-aided design) drawing created by Christopher L. Johnson, an Airport Planner. With degrees from the University of Michigan Taubman School of Architecture + Urban Planning, in both Architecture and Urban Planning, Mr. Johnson designs airport facilities and buildings to accommodate current and future air travel changes. flight patterns, sizes of aircraft, climate, the needs of surrounding residential and business communities impacted by air travel, and a myriad of other variables that may change over time. Mr. Johnson helps ensure that airports stay safe in the midst of change, and that they keep up with the times. Just as his important endeavor, his niche, combines skills in architecture, urban planning, engineering, business and community relations, so might the opportunities that your diverse students experience as they combine various areas of their training and expertise. Help them explore opportunities that engage their talents and interests for future fulfillment.

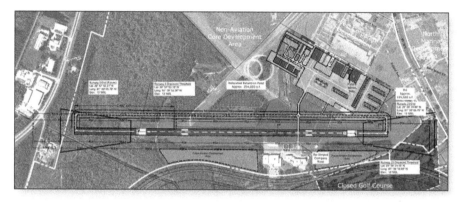

Conceptual Florida Airport Layout Plan (ALP) created by and Courtesy of Christopher L. Johnson, Airport Planner, Passero Associates, LLC, 2022.

***

The A2GO is an autonomous (self-driving) shuttle at the University of Michigan. It draws hundreds of riders every day since it began operation in the fall of 2021. This on-demand autonomous (driverless) shuttle service is free to the public, and connects the University of Michigan campus, Kerrytown, and the near campus State Street corridor in Ann Arbor, Michigan. As noted by the University of Michigan Mcity (collaborative initiative transforming mobility for society), the ratings of riders have been high, 4.81 out of 5 stars. The A2GO fleet includes five autonomous on-demand vehicles, four hybrid-electric Lexus RX 45 vehicles that carry up to three passengers, and one Polaris GEM fully electric vehicle that has capacity for one passenger in a wheelchair. Rides are ordered from a free mobility app available through Apple's App Store or Google Play. A2GO partners include Ann Arbor Spark, Important Safety Technologies, the State of Michigan Office of Future Mobility and Electrification, and the Michigan Economic Development Corporation (MEDC).

As these innovations occur, they bring many opportunities for diverse students. What an exciting time to engage our children in all of the STEM developments taking place!

Photo and background courtesy of Susan Carney, Communications Director of Marketing and Communications at the University of Michigan Mcity. 10/21.

***

At Michigan State University in East Lansing, Michigan, the ADASTEC Bus drives through campus. ADASTEC, an automation solutions company, described the vehicle as an electric Level-4 autonomous bus, developed in partnership with Michigan State University and the State of Michigan mobility grant program. Detroit, Michigan, the historic "Motor City" and home of transportation and the future of mobility, will be the headquarters for ADASTEC, a San

Francisco-based company delivering advanced automated transportation platforms for full-size commercial vehicles.

Self-driving vehicles, commercial and domestic, are the wave of the future, and many STEM professionals, including diverse students and scientists of color, will be leading the charge in these developments.

Photo and background information courtesy of the Michigan
Economic Development Corporation, 2022

***

Also at Michigan State University is a physical science marvel, the facility for rare isotope beams, or the FRIB, which processes quantum information. FRIB, a 100,000-pound "coldbox" used to cool liquid helium to 60 degrees Kelvin, is now available for people to come in and observe. MSU built the FRIB's predecessor, the K-50 Cyclotron, a national superconducting cyclotron laboratory, in 1960. The nuclear physics program at Michigan State University has trained many nuclear physicists from across the U.S. Data on the decay of nuclei gathered at the FRIB may provide information about how to destroy nuclear waste that otherwise will sit around and affect mankind for eons to come. Diverse students enrolled in these areas of science may one day assist in the many ways that the FRIB and other physics laboratories will aid mankind.

***

As we look around us, so many opportunities exist for our children to immerse their talents in STEM/STEAM careers!

As many leaders gather internationally to discuss climate change, so are these meetings taking place in the Great Lakes states. A recent discussion included the Director of the United Native American Tribes of Michigan, professors from York University, and the Michigan Department of Natural Resources, Fisheries Division. It is important that tribal communities engage in these talks, and are empowered and educated to take leadership in how Michigan and other Great Lakes states address climate change. United Tribes of Michigan Executive Director, Frank Ettawageshik, is a citizen of the Little Traverse Bay Bands of Odawa Indians. The United tribes of Michigan's mission is: "Committed to joining forces, advance, protect, preserve and enhance the mutual interest, treaty rights, sovereignty and cultural way of life of the sovereign tribes of Michigan throughout the next 7 generations." This commitment will also include the educating of Native American children and families, empowering tribal communities with resources to help advance tribal youth in STEM/STEAM careers.

As we look at the Indigenous populations that helped define America, we are aware of all of the wonderful contributions they have made to this nation and great land. Collaborative discussions take place about indigenous futures, and how we all work together to sustain, through climate change. Also, shifts in where and how people live, how we will ensure equity for all people in our nation, African American, Latino and Native people, and especially how we will equitably educate future generations.

Professor of Psychology Stephanie Fryberg is a member of
Tulalip Tribes of Washington and the Director of RISE.
*Photo by Sandy Chapin, University of Michigan LSA Magazine,
Summer 2022, Indigenous Futures, by Susan Hutton.*

***

Immediate opportunities also exist for our future STEM professionals in atmospheric sciences. In its Geospace Dynamics Constellation Mission (GDC), NASA took a first time look at a protective outer layer of Earth's atmosphere and the way it interacts with solar weather. Researchers have noted that solar weather poses a threat to earth, with the potential to cause major damage to our power grids and satellites. The NASA GDC is doing scientific investigations that will help predict impacts from solar wind and sun flares. This will be an ongoing discussion as we consider how climate change and global warming is affecting transportation in Europe, and our infrastructure worldwide.

Additionally, in the areas of equipment and cyber security, many STEM career opportunities will exist in protecting our data. IT careers will be important in helping secure technological services and electronic devices. As artificial intelligence (AI) increases, a STEM workforce to provide safe data access will also be needed. Many schools, colleges, and career programs are training youth for these roles.

Certainly, it is important for families, parents, and the community, to encourage diverse students, African American, Hispanic, Native American, Asian, Middle Eastern, women, and first-generation students in emerging STEM opportunities.

In the world of oceanic sciences, as in many others, there are great discoveries for STEM students of color, first-generation and female students to enjoy. Consider how humans can recognize each other by combining several traits, including a person's gait, the sound of their voice, their laugh, hair appearance, or eyeglasses. However, how do we explain how dolphins can use multiple clues to zero in on friends, such as their unique whistles or their taste. Scientists discovered that dolphins paid more attention to their friends' whistles than those of strangers, suggesting they knew the animals that issued them. With so many scientific marvels to enjoy, this is a great time to be a STEM student or professional in aquatic sciences.

And sharks, we are learning more about these deep-sea creatures. National Geographic's 10th anniversary event, entitled "SharkFest," features videos about how great white sharks may change their color to sneak up on prey. Are your students interested in oceanographic careers and learning more about such discoveries as these and many others that they can take part in?

Then there is deep space. NASA has released the first images from the new James Webb Space Telescope (JWST) that may change the way scientists characterize outer space. According to NASA, soon to be launched images from the 3D modeling telescope will enable astronomers to study star birth, star death, galaxy formation from the first galaxies across cosmic time, and planets around nearby stars. Information courtesy of UM Astronomy Professor Michael Meyer, UM Michigan News, July 12, 2022.

Is Astronomy and space science in your children's STEM future?

Speaking of NASA and space exploration careers in which your children might have interest. Many saw the movie "Hidden Figures," released in December 2016, about three brilliant African American women at NASA, Katherine Johnson, Dorothy Vaughan, and Mary Jackson, who helped to transform operations at NASA in their respective fields, and served as the brains behind the historic and successful launch of Astronaut John Glenn, the first American to orbit the earth. That event was heralded around the world as America took the lead in space travel and exploration. Prior launch attempts had failed, and the movie depicted how mathematician, Katherine Johnson, created the precise trajectory calculations needed to successfully and safely launch Astronaut John Glenn for the three-orbit, Mercury-Atlas 6 mission on February 20, 1962, on the spacecraft Astronaut Glenn named the Friendship 7.

In addition, Katherine Johnson worked on the Apollo and space shuttle programs. Mary Jackson was the first African American female in aeronautical engineering in the 1950's, who aided space explorations for many years. Dorothy Vaughan was NASA's first African American staff Supervisor.

The movie was based on the book, "Hidden Figures: The American Dream and the Untold Story of the Black Women Mathematicians Who Helped Win the Space Race," written by Margot Lee Shetterly. She grew up in Hampton, VA, and is the daughter of a former NASA/Langley scientist. The book and subsequent movie shed light on these significant STEM heroes in American space travel who might have remained unsung heroes. They are "hidden figures" no more. Perhaps your students will be inspired by the stories of these brave and brilliant women of color. There is much STEM discovery ahead in the explorations of space!

(L to R) Dorothy Vaughan, Katherine Johnson, and Mary Jackson
Photo Courtesy of Communications ACM, February 2022 (Vol. 65, No. 2)

(L to R) Mary Jackson (portrayed in the movie "Hidden Figures" by Janelle Monae'), Katherine Johnson (portrayed by Teraji P. Henson), and Dorothy Vaughan (portrayed by Octavia Spencer). Photo courtesy of Astrobites.com 2016

Mary Jackson was the first African American female
in aeronautical engineering in the 1950s.
Photo by Bob Nye, Courtesy of NASA/Langley, npr.com 12-16-16

Katherine Johnson, NASA mathematician, sits at her desk with a globe, or "Celestial Training Device." Photo Courtesy of NASA/Langley, npr.com 9-25-16

***

As well as, we speak of women in STEM and space science, Astronomer Cecilia Payne is a hero who discovered hydrogen is the most abundant element in the universe. Dr. Payne passed away in 1979 but never received recognition for her greatest discovery. As noted by Jeremy Knowles, Schlesinger Library, "Every high school student knows that Isaac Newton discovered gravity, that Charles Darwin discovered the theory of evolution, and that Albert Einstein discovered the relativity of time. However, when it comes to the composition of our universe, the textbooks simply say that the most abundant atom in the universe is hydrogen. And no one ever wondered how we knew this." Jeremy Knowles shared for posterity that Dr. Cecilia Payne deserves this scientific acknowledgement.

In her youth, it was not popular for a woman to become a space scientist. Dr. Payne's mother and family did not assist in financing her education, but she won a scholarship to Cambridge University in England. Cambridge would not give her a degree after she completed her studies because she was a woman, so she transferred to Harvard University in the U.S. and was the first person to be granted a PhD in Astronomy from Radcliffe College at Harvard. According to Otto Strauve, Dr. Payne's PhD thesis was the most brilliant ever written in astronomy. He said, "Not only did Cecilia Payne discover what the universe is made of, she also discovered what the sun is made of. Henry Norris Russell, a fellow astronomer, is usually given credit for discovering that the sun's composition is different from the earth, but he came to his conclusions four years later than Payne, after telling her not to publish her findings".

Dr. Cecilia Payne is the reason we know about variable stars, that is stars whose brightness is seen from earth to fluctuate, or 'twinkle'. Literally, every other study on variable stars is based on her work. Cecilia Payne was the first woman to be promoted to full professor from within Harvard, and is often credited with breaking the glass ceiling for women in the Harvard science department and in astronomy, as well as inspiring generations of women to aspire to scientific careers. Dr. Cecilia Payne's discoveries are a gift to science, as are the great contributions of many other brilliant women in STEM, as we have noted. Going forward, many diverse students, women and first-generation students can be inspired by Dr. Payne's legacy.

Dr. Cecilia Payne
Photograph courtesy of Schlesinger Library and Jeremy Knowles, 4-23-22

***

As well as, for future medical scientists there are amazing innovations, from creating implantable body parts, using 3-D printers, to seeing life-like digital images of the human body through your laptop computer. Applications of 3D imaging of anatomy and physiology can be fully explored on a computer from anywhere. Not so long ago, the study of anatomy and physiology could only be done in person, in labs. Your children will be able to explore these and many more innovations through their science programs at school, but also through accessible and innovative science applications, camps and workshops sponsored by schools, colleges and scientific organizations.

***

Parents Make the Difference

REMEMBER THAT AS a parent, you are enough just as you are. Build a great relationship with your children that will last a lifetime. Build it and strengthen it. Praise them when they do well, and share what makes you proud. It is great to say to your child from time to time, you are awesome!

Designed and created by Fredrick Johnson, Design Artist and Illustrator, 2022

***

"You must be the change you wish to see in the world."

Mahatma Gandhi

In the inspirational movie—"The Real MVP: the Wanda Durant Story", starring Sandra Freeman, NBA basketball player, Kevin Durant tells about his mother, the amazing parent he says guided him and his brother, Tony Durant, to successful fulfillment of their dreams. It shows how determination leads to positive outcomes. Durant played one season in college with the Texas Longhorns, was the second pick by the Seattle Super Sonics in 2007, and he currently plays for the Brooklyn Nets. This is one of many stories we hear about loving and supportive parents making the difference in one's life.

In Hollywood, many actors and artists talk about the importance of their parents and how they helped shape their lives on stage and screen. In the May 30, 2022, (issue 22), US Magazine Weekly, noted the mothers of Drake, Elon Musk, Pete Davidson, Leonardo DiCaprio, Bradley Cooper, Chris Evans and others, discussing how they helped their children hone their craft. In addition, many fathers have shared their pride, as the father of performing artist H.E.R. It is clear that parents played a major role in helping them present their talents to the world.

Your full presence is the greatest gift you can give your children in this moment. Anchor yourself in the now and give them your all. Notice the details as they grow and develop. Be present. As you get to know your talented children, nurture their vision, their unique gifts are meant to be shared.

Besides, that brings us to the story: **The Life and Times of Chicken Charlie**

The Science of Compassion We Teach Our Children

Do you and your children have memorable pet stories? Stories they may recount many years from now, the joy of bonding with animals, compassion, and the sense of wonder? On the other hand, just feeling awesome remembering the experience, for me, is the story of my beloved pet chickens, Charlie, Larry, Fred and Sam, and particularly the heroic (Drum Roll please...) **Chicken Charlie!**

Charlie will forever be remembered as one of my heroes. He far exceeded any expectation we generally attribute to pets or farm animals. He exemplified strength and endurance, and refused to let his unexpected, life-altering accident get the best of him or stop him from proceeding on his path. Perhaps Charlie can inspire our children, and adults as well! Nevertheless, I am getting ahead of myself.

Illustrations by Fredrick Johnson, Graphic Artist, 2022

The story begins on a spring day, just before Easter, when I was 8 years old. My mother and I were Easter shopping at Spartan Atlantic, a department store at that time, in Inkster, MI. As we left the store, we noticed farmers placing stands in the parking lot, displaying crates of beautiful, chirping, yellow baby chickens. We walked over to observe them and my mother asked if I would enjoy having some 'real life' Easter Chicks this year. Delightedly, jumping up and down, I exclaimed, Yes! As well as, that day I met four amazing friends, albeit baby chickens, that I will forever remember. As my mother drove us to a farm store, the chicks nestled in their crate on the back seat; mom explained that she knew nothing about raising backyard chickens, so we would be learning together. A kind farmer walked us around the store and pointed out the items we would need—wood and structures to build a chicken coop with attached pen, bedding, chicken feed, food storage container, feeders, waterers, treats and supplements.

Gathered with supplies, we arrived home and introduced the chickens to their new residence in our backyard. It was such an exciting moment! As we built their refuge and provided them their first feast of chicken feed, water and treats, the four baby chicks nestled around our feet and seemed right at home. Over the next several months, the wee wonders grew larger and heartier...and they began developing red cones on their heads and under their beaks. We had been unsure of their gender at first, but it was apparent now that we had ourselves some handsome bantam roosters! At about 3 months old, they began crowing every morning at 5 or 6 a.m., announcing to the world, 'Good Morning everyone, we're here!' Though this startled our not so thrilled neighbors, and us, I loved our pet roosters too much to mind. Mom smiled and said, well, it is time to name our little friends, and right away I came up with—Fred, Sam, Larry and Charlie.

I became so attached to them, that they seemed more like people than chickens. They even tried to follow me to school, but my mother called them back, and they waited each day for me to appear around the corner, so they could accompany me home.

One fateful day, I went to the back yard to feed them, and as I returned to the house, Charlie, unbeknownst to me, attempted to run in the screen door with me, and the door accidently hit him. When I heard his loud, mournful cry, I tearfully looked down to see Charlie's leg was broken, it had been caught in the closing door. I ran frantically to my mother, praying there was something she could do to save Charlie. Mom, the consummate nurse, calmly and compassionately laid Charlie on a palette, calmed him from trembling, and began determining what we could do to address his painfully broken leg. After calling a farm veterinarian, my mother quickly treated and bound his upper leg with a tiny "ace" bandage she fashioned, gave him penicillin in his water to prevent infection, and then she and I hugged and nurtured brave little Charlie over the days until he was better. Soon the veterinarian said it was all right to create a prosthetic leg for him, so my mother got a dowel, sterilized it, and attached it skillfully with bandaging to help Charlie walk again. Before we knew it, Charlie was up and running around the yard on his new peg leg.

Illustrations by Fredrick Johnson, Graphic Artist, 2022

Charlie wasted no time learning how to keep up with his brothers, and he actually seemed to outrun them when they came to greet me after school. Such was the resilience of this amazing and loyal little friend, my Chicken Charlie! One day, coming home from school some kids chased me as I rounded the corner. When I looked up, all four of my powerful roosters had come running, some flying, wings spread, to protect me. They lit into my assailants with pecking force that sent them all screaming frantically away. Needless to say, they never bothered me again. Additionally, what was most amazing was how Charlie seemed to lead the charge, with force and grace, running stronger and faster than the rest. He was my fearsome warrior, my hero, my friend. Apparently, he never held it against us that his leg had been permanently transfigured, for every day he seemed happy just being a part of the family we had become. He seemed to walk taller, knowing that everyone praised his strength of being a survivor and a hero. I believe Charlie will forever be an inspiration to all of us who had the privilege of knowing him, and for all who read this story.

As the years have gone by, any time that loved ones or I have experienced illness or injury in life, I think about my Chicken Charlie. His bravery and determination to get up and go despite it all, gives a lesson to remember on love and endurance.

Illustrations by Fredrick Johnson, Graphic Artist, 2022

I share this story as an example of ways that parents and children learn together in STEM/STEAM experiences. My mother and I learned about veterinary medicine and animal care in ways we never would have otherwise. Perhaps this will inspire your children as they develop interest in animal science.

Children learn much through sharing and engaging with parents. Whether you are raising beloved pets, tending a garden, or learning about planting, composting and soil enrichment. Or repairing an automobile, bicycle, motorcycle, lawn mower, snow blower, or other mechanical devices, children will learn about STEM sciences in various forms from you. Continue to share your many skills and talents, and joyous adventures, with your children.

Culinary times are also fun. As they help you cook, bake, marinate, smoke foods, or grill favorite dishes and confections, your children will learn food science and the physical processes of cooking. As the family gathers together over the preparation of a family favorite—like Aunt Annie's Southern Pound Cake, Uncle Carl's West African Jollof Rice, Elder Walking Eagle's Lakota Plum Cake, or Mi Abuela Adelina's Sopapillas—parents can tell rich family stories of native heritage, anchoring in the love of oral history, and the pride of their culture.

When your children go fishing or hunting with you, they will learn about environmental, aquatic, and animal science, not to mention meteorology, especially when a sudden thunderstorm or snow squall catches everyone by surprise and sends you sprinting for safety, or you observe the heavens, and there is a shooting star. You can share knowledge about aquatic life, invasive species, zebra and quagga mussels for example, in the Great Lakes region, or about the Asian Carp, a species causing concern. There are so many world and environmental discoveries. Your knowledge and willingness to inform and share curiosity with your children will inspire great conversations and stories, many of which can be teachable moments for them, as they become STEM professionals.

"Every Child is an Artist..."

Pablo Picasso

As parents, you have the opportunity to acquaint your children with STEM/STEAM mentors who may continue to be a blessing in their lives years later, and render stories they will cherish.

In my hometown of Inkster, Michigan, my mother took me when I was a small child to the home of dear friends, Godfrey and Della

Jones, members of our Womack Temple C.M.E. Church family. There I met their daughter, the wonderful Gladys Horton, who was forming a singing group of friends for the Inkster High School talent show. At times during my "babysitting visits" I would hear them rehearsing, not knowing that one day, they would become legends. The original members were Gladys C. Horton, Katherine Anderson, Georgeanna Tillman, Juanita Cowart, and Georgia Dobbins. Wanda Young later replaced Georgeanna Dobbins in 1961. An Inkster High School teacher, Ms. Shirley Sharpley, and her colleagues, facilitated the group meeting the Detroit Motown Records Producer, Berry Gordy. He liked their sound, encouraged them to change their name from the Casinyets (aka "Can't Sing Yets"), to the Marvelettes, and signed them to the Tamla/Motown label. Their first song, "Please Mr. Postman," led by Gladys Horton, became the recording that put Motown music "on the map," and on the international sound stage, loved by people and cultures around the world. It became a #1 U.S., International, and Rock and Roll hit, and the first number one single at that time to be recorded by any all-female or Motown recording group.

This Marvelettes debut song has since been remixed, sampled, and covered more than 10 times by a number of artists, including The Beatles, the Carpenters, and the Muzak Orchestra. The Marvelettes went on to record many more top hits. Ms. Horton took time away to raise her family, and then returned to performing.

Years later as an adult I had the serendipitous opportunity to reconnect with Gladys Horton through colleagues in California. The phone call that day from my mentor, Gladys Horton, is a reunion I will never forget. I was so honored to speak with her, those many years later, and update her on my life, career, and family experiences since I had last seen her. She in turn shared remarkable stories of her amazing musical career, her beautiful family and the life and trajectory she had enjoyed as a renowned musical artist.

In 1999, the opportunity of a lifetime came when Gladys invited me to join her on a road tour as a back-up "Marvelettes" singer, and she said I could bring another singer to join us, one with whom I could rehearse prior to the tour. I selected my talented daughter, Dichondra! Those 10 days on the road, performing with a Motown Legend and consummate entertainer, Gladys Horton, changed my life. Besides, as a parent, I was so proud that I could share this wonderful experience in the arts with my daughter. Ms. Horton gave us a unique STEM and Arts experience, quickly mentoring us in professional musical performing, interactive technology, and the business side and acumen of managing live performances successfully.

Reflecting on this story, reminds me that as parents, we never know whom we will meet, or when opportunities and experiences will come that we can share with our children.

As her parents, mentors, teachers and community nurtured Gladys Horton, helping her ascend to musical greatness, she has bountifully inspired many around the world as a remarkable and talented artist. The book that she authored, "A Letter from the Postman: A Memoir by the Original Lead Singer of the "Marvelettes," is a must-read tribute to her legacy. Published in 2022, it is a keepsake for all music lovers and loyal fans around the world to enjoy.

Gladys Horton will forever be treasured as my mentor and friend, and an inspiration to young and aspiring STEM/STEAM artists, singers, music writers, technologists, and producers everywhere. I thank her son, Vaughn Thornton, for carrying forth his mother's legacy, and for the permission to share this personal tribute to a great artist, phenomenal woman, and an amazing parent! Thank you always and forever, Gladys!

Mid 1960's photo of The Original Marvelettes founded by Gladys Horton, (left to right) Gladys Horton, Wanda Young, Georgeanna Tilman and Katherine Anderson. Photo Courtesy of Giles Petard/Redferns, 2011.

Image Courtesy of Vaughn Thornton, 2022

GLADYS HORTON

Gladys Horton, photo Courtesy of Vaughn Thornton, 2022

"The future belongs to those who believe in their dreams."

Eleanor Roosevelt

Graduates, All!

"Won't it be wonderful when black history and Native
American history and Jewish history and all of U.S.
history is taught from one book. Just U.S. history."

Maya Angelou

ACADEMIC ACHIEVEMENT IS important in all cultures. In
addition to African Americans and Native Americans, many Latino
and Hispanic Americans, Asian Americans, Arab Americans and
Pacific Islanders all experience challenges as they work toward educational and career attainment.

As Michelle Obama stated in a commencement message: "Parents
of all cultures and ethnicities are called to action to ensure support
and encouragement for their students, and to partner with all educational personnel, agencies and resources you identify to aid in this
support. In some situations, you will need to ensure English transla-
tors or tutors to gain language proficiency. You will need financial aid
for your students' academic success. Moreover, everyone will need
technological access in this 21st century – including internet capa-
bility, computers or smart devices, and continual means for electronic
communication. Although cultures are different, these resources are
universal and fundamental for helping you and your children learn,
connect to schools, activities, the community, and the world."

Michelle Obama's Message to Native American Students
Story and photo Courtesy of nbcnews.com, 2022

***

During the global pandemic, those who expressed stress in keeping up, from all cultures, said that poor or non-existent Wi-Fi and equipment, was the bane of their existence. They could not communicate, or complete their courses or professional work in a meaningful and timely manner. The good news is that many schools and colleges have begun offering computer equipment and Wi-Fi options upon request for parents and students to rent or borrow. Thank goodness that even in a global crisis, technology enables education to continue.

Congratulations to the graduates!
Photo Courtesy of the American Indian College Fund, 2022

President Barack Obama, flanked by former Michigan Governor Jennifer
Granholm on the left and former University of Michigan President Mary
Sue Coleman on the right, greets graduates and the crowd of more
than 80,000 at Michigan stadium in 2010. Story and Photo courtesy
of the Ann Arbor News, Ann Arbor Michigan, January 23, 2012.

Arizona law now allows Native American graduates to wear tribal regalia. Kirtland Central High School graduate Angel Yellowman wears regalia from the Cheyenne Sioux River during the May 2021 commencement ceremony in Kirtland, AZ. Story and Photo Courtesy of Noel Lyn Smith, The Daily Times, Kirtland, Arizona, May 16, 2022

This is a Latina and Hispanic perspective to which many parents and students can relate.

As Wendy Medina stated in her article "First-Gen Latinas Talk College Experiences: Realities, Pressures, Successes" in a recent mitu' segment, "Pursuing higher education is no joke, a dream unattainable by some and grueling for those who decide to attend. Many emotions come with the college experience for Latinos born in this country to immigrant parents, from pride and joy when receiving that spring acceptance letter, to guilt and impostor syndrome when it's time to walk that commencement stage and long after period." Ms. Medina states that Latinos face different challenges, having to be a translator and an advisor as a 5-year-old, navigating a foreign speaking world with no prior knowledge, reaching heights that surpassed her ancestor's wildest dreams. She says "However, it isn't always sun-shine and rainbows. After making it, many of us are still torn between building our own independent lives or opting out of higher education or fulfilling careers to support our families." There is a sense of respon-sibility that she and others face to help family members who are not pursuing careers or college education. She shared how her parents dropped out of school during their elementary years. Therefore, to them, her finishing high school was a great accomplishment. When she is around college friends who she feels are from privileged back-grounds, she does not relate as well to their conversations about their school experiences, since her challenges are very different from theirs. She said, "I was out of luck and had to figure things out on my own, *me puse las pilas*".

She did not have a car, so she had to get up at 5:00 a.m. to take the bus for her 11 a.m. class. She wanted to get to campus early so she could use the computer lab before class, until she was able to purchase her own laptop in her senior year. She is no longer in col-lege and is now working as a professional, but still feels the pressure as the first-generation graduate in her family, and with this pressure, unfortunately comes the guilt. She feels guilty for working from home while her parents are out doing manual labor. She spends time with them when she can and takes them for opportunities she knew they did not get to experience when they were younger. Being a college educated professional has given her perspective, and has helped her be a better and more empathetic communicator, with her family and in her life. Wendy said, "I hope my parents are proud of me, because I know I am proud of them."

Pictured below are fellow graduates.

From left to right, images Courtesy of Anggye D. Godoy Garcia, Audrey Diaz Robles, Maria Alonso, Valerie Romero Rosas, in wearemitu.com, 2022

Latina college graduate, Anna Ocegueda, gives tribute to her immigrant farm worker parents. Story by Ludwig Hurtado, nbcnews. com, May 15, 2019. Photo Courtesy of Anna Ocegueda.

By Yesenia Robles, Reporter, Chalkbeat Colorado, June 2, 2022
Marisa Beltran, who graduated high school in 2015, was part of a
decade-long trend of increased Hispanic high school graduation rates.
Her mother Rosa, who dropped out as a teenager, always encouraged
Marisa to graduate and attend college. Driving a decade of progress,
Hispanic students made huge gains in high school graduation.
Photo Courtesy of Carl Glenn Payne II for Chalkbeat, Colorado, 6-2-22

Photo Courtesy of Drexel University, Downsife School of Public Health, 2022

Northwestern University Commencement, Class of 2022.
Photo Courtesy of Northwestern University, Evanston, IL, May 2022

Purdue University winter 2021 commencement, December 17, 2021
Photo courtesy of @LifeAtPurdue, Purdue University, West Lafayette, IN

College Graduates sharing hope and excitement.

Kent State University graduates, May 20, 2016
Photo Courtesy of Kent State University, Kent, Ohio

Southern Oregon University Commencement Ceremony

Doctoral Commencement Photo of Dominic Bednar, PhD,
Spring 2022, University of Michigan, Ann Arbor, MI
Photo Courtesy of Karl Paine Photography & Design, 2022

University of Michigan Commencement, Spring
2022. Celebrated for the first time since 2019.
Photo Courtesy of the University of Michigan, www.umich.edu, Ann Arbor, MI.

Wayne State University Graduation Ceremony, Detroit, MI, April 2022
Photo Courtesy of Wayne State University, Detroit, MI

In May of each year, the Harvard Arab Alumni Association (HAAA) organizes a graduation ceremony for Arab students and their families to commemorate their academic achievements and wish them well. Photo Courtesy of the Harvard University HAAA, May 2017.

Muslim student, Hina Haider, a Chancellor's Medallion recipient, who received a B.A. with High Distinction, graduating with a 4.0 GPA from the University of Michigan-Dearborn, May 2, 2016. Story and Photo Courtesy of University of Michigan-Dearborn, Michigan.

Michigan State University Race and Ethnicity students, participate in a diversity study abroad project. Photo Courtesy of MSU International Studies & Programs, Office for Education Abroad, 2022

Lansing Community College Commencement, Spring Semester 2022
Photo Courtesy of Lansing Community College, www.lcc.edu, Lansing, MI

Ferris State University Commencement, May 7, 2022
Photo Courtesy of Ferris State University, www.Ferris.edu, Big Rapids, MI

Eastern Michigan University Commencement, Spring 2020
Photo Courtesy of Eastern Michigan University, www.emich.edu, Ypsilanti, MI

Western Michigan University Commencement, Spring 2021
Photo Courtesy of Western Michigan University,
www.wmich.edu, Kalamazoo, MI

The first cohort of students has graduated from the Detroit
Apple Developer Academy in partnership with Michigan
State University. https://www.go.MSU.edu/Dgp5

Apple, Michigan State University, and the Gilbert Family Foundation celebrated the accomplishments of graduates from the first cohort of the Detroit Apple Developer Academy in summer 2022.

The academy, located in Detroit, Michigan, is the first in the U.S., launched as part of Apple's Racial Equity and Justice Initiative. The program is free of charge and offers students an opportunity to become entrepreneurs as well as app developers. They learn the fundamentals of coding, design, marketing and project management — with an emphasis on inclusivity and making a positive impact in local communities.

The academy is already recruiting for its next cohort of students and is accepting applications for the upcoming class on a rolling basis. Individuals 18 or older from all backgrounds who have an interest in building a foundation for a career in app economy are encouraged to apply. Enrollment is available at no cost, and students are not required to have any previous coding experience. Students in this year's class bring a breadth of personal, professional, and academic experience to the program.

For more information on the programs and ways to get involved, visit the Apple Developer Academy.

Information and photo Courtesy of Michigan State University and the Apple Developer Academy, 2022
Developer academy.MSU.edu

Central Michigan University Commencement, 2021
Photo Courtesy of Central Michigan University,
www.cmich.edu, Mount Pleasant, MI

Marianne Williamson's beautiful poem, "Our Deepest Fear"
Also loved and recited by President Nelson Mandela

"Our deepest fear is not that we are inadequate.
Our deepest fear is that we are powerful beyond measure.
It is our light, not our darkness
That most frightens us.

We ask ourselves
Who am I to be brilliant, gorgeous, talented, fabulous?
Actually, who are you not to be?
You are a child of God.

Your playing small
Does not serve the world.
There's nothing enlightened about shrinking
So that other people won't feel insecure around you.

We are all meant to shine,
As children do,
We were born to make manifest
The glory of God that is within us.

It's not just in some of us,
It's in everyone.

And as we let our own light shine,
We unconsciously give other people
Permission to do the same.
As we're liberated from our own fear,
Our presence automatically liberates others.

Understanding and Cultivating Social Capital

"Knowledge is power, community is strength,
and positive attitude is everything."

Lance Armstrong

WE KNOW THAT in STEM education and careers, people—teachers, counselors, mentors, and influential friends—play a huge role in a child's social and academic success. These people can and will open doors of opportunity for your children. Students must learn early about the importance of cultivating relationships through developing a common ground, a sense that they can communicate easily with anyone, and care about the other individuals that they meet. In addition, children encounter different socioeconomic classes at school and in society. They will benefit from knowing how other people in their academic environments think and express themselves. The children, parents, families with whom they engage, their social and cultural networks, all present opportunities for growth through interaction. Learning to understand the different classes and cultures of people will help them become engaging communicators with other students, faculty, and those in the STEM community with whom they will work.

Dr. Rudy K. Payne, author of "*A Framework for Understanding Poverty*," 1995, has written extensively about how having what Dr. Payne defines as "social capital" or "social register" aids in a student's ability to communicate broadly as they proceed in academia. Some would call this ability social intelligence, or mother wit. Others call it charm or charisma, or "he or she never meets a stranger" kind of skill. It is important to have, or develop, a comfort level in communicating with all classes of people. Empathy, understanding, and open-mindedness are necessary to engage the many people that we and our students will meet.

Understanding the Social Classes

Given humanity's different cultures, lifestyles and experiences, a part of diverse awareness is to understand the 'social capital', or the socio-economic class expression of your children's educators, counselors and mentors. Dr. Payne says that socioeconomic 'class' experiences affect how people think, speak, experience life and express themselves. This difference can affect how educators work within culturally sensitive situations with students, particularly if they have different socioeconomic backgrounds.

It is important for diverse parents and students of color, regardless of their ethnic or socioeconomic backgrounds, to communicate effectively and confidently, to understand and be understood, by expressing and conversing observantly. Being good communicators, in school and in life, aids students' ability to succeed academically and secure their places in STEM/STEAM programs and opportunities.

How Social Classes Communicate

Dr. Payne stated that good communicators, and the society at large, typically use a "formal language" that features complete sentences and specific word choices, depending on the topic of conversation. Casual language, or casual 'social register' which, according to Dr. Payne, includes broken sentences and non-verbal gestures, are not generally the best way to communicate with educators. She indicates that students using these forms of communication can suffer academically because they do not communicate in the "language" that their teachers utilize and understand. She provided the following example. When a particular student was asked by the teacher why he did not turn in his math assignment, he responded to the teacher the assignment was "whack", "didn't make sense", and "that's why I didn't complete it". This student received a reprimand from the teacher. When the teacher asked another student why he did not turn in the same assignment, the student replied of the assignment "I did not follow the earlier instructions that you gave us, because I didn't understand them, and I was unsure of what to do." This student was not reprimanded and was encouraged to stay after class to receive assistance. Yet both students had expressed the same concern, that they did not understand the assignment. In this example, the first student was speaking in what Dr. Payne called 'casual register', while the other student was speaking in 'formal register'. This example shows the importance of thinking about communication when speaking with their teachers and mentors, with the goal of fully expressing what they need while in the school environment. Guide your students to always take time to communicate in detail, fully and intelligently express their questions and concerns, and use full sentences, not 'casual' slogans or gestures they might otherwise use among their peers.

Let us now revisit this example, building upon recommended social etiquette. The first student could, in 'formal register', raise his or her hand, and state to the teacher: "Mr. or Ms. _____, in completing this math assignment, I am attempting to understand the entire equation, and I am unsure about the calculations after the first step. Would you explain the problem again, please, so that I can get the remaining steps? I am happy to meet with you after class if necessary." Now the question has been fully and properly conveyed, and the instructor can respond in kind with help for the student. If this were your student, and he or she continued to need assistance after formally addressing the teacher, then you could reach out as a concerned parent, to further discuss and engage the teacher's assistance for your child. That is the support that aids, when needed.

We, as diverse parents, can recommend options that bring positive, socially healthy outcomes for our children, helping them to be aware, and to hone their academic and social communication skills.

***

"Just don't give up what you're trying to do. Where there is love and inspiration, I don't think you can go wrong".
                                                                    Ella Fitzgerald

Communicating and Learning

AS HAS BEEN said, despite the realities of bias in this nation, it is clear that parents, regardless of their past circumstances, can significantly aid the education, careers, and legacies of their children.

Unspoken class and cultural barriers should not hinder our ability as parents to feel comfortable around our children's various learning environments, whether it be with the personnel at their early childhood development centers, K-12 schools, college faculty and administrators, or educators at STEM technology training centers.

Remember, you define your children's world. We know that they want, need, and deserve love, care, affection, and attention. Our children are a gift, a heritage, and part of our legacy and journey. Children count on parents to help them carve their path in society. We as parents do not have to understand the intricacies of STEM in order to put our children on the path of opportunities that could lead them to their STEM/STEAM potential.

This primer seeks to encourage you to inspire and mentor your children in STEM. Here are a number of areas to aid your quest.

Early Start Practices to Aid Children's Development in STEM

Research has shown that children who have parents and families involved in their academic lives get better grades and test scores, graduate from high school at higher rates, are more likely to go on to higher education, and are better behaved and have more positive attitudes. Children will be competitive in STEM/STEAM careers if they are properly prepared, and it is never too early to start guiding them. You can inspire your children's learning at every age, even when they are still in the womb. The following list includes many ways to encourage your children to love learning at any age.

Language and Reading

As noted in my doctoral research (*V. Wilkerson-Johnson, 2015)* you can help your children develop a love for learning, literature, science, and math by reading, talking, and singing to them during pregnancy. By the 6th month or third trimester of pregnancy, the baby's brain and auditory system are developed enough to hear and recognize sounds. Babies in the last trimester can remember nursery rhymes that mothers, fathers, siblings, and other family members read to them while they were in the womb. Experts agree that if you continue to read the same book after birth, the baby responds to the story by recognizing the sights and sounds. The practice of reading to your baby in utero is also a nurturing experience, providing a bonding opportunity between a mother and child. Creative expression, imagery, the power of language, and the sequencing of numbers are all learned readily when children are nurtured and read to in utero and after birth.

In their book *Can't Wait To Show You: A Celebration for Mothers-to-Be*, 2014, Boyle and Stonis provide suggestions for reading to children before and after birth. They encourage pregnant women to rest the book on their bellies, reading to the baby slowly, clearly, and audibly. By relaxing and enjoying the language and imagery, mothers will improve their emotional connection with the baby each time they read. Furthermore, mothers can sing some of the verses like a lullaby during times of reading, as newborns recognize a song that is heard repeatedly from inside the womb and are calmed by a melody sung by a familiar voice.

Then, continue to read to and with your children as they grow. Children from infant to five years of age benefit from early learning centers. In addition, children will benefit from daily reading, age-appropriate math toys, letter and word games, and age-appropriate music and speech recordings to stimulate their joy of learning. As has been noted, children who read at home with their parents perform better in school than those who do not. There are many ways you can encourage your children to read, such as by keeping good books, magazines, and newspapers in the house. Also, let your children see you reading. Take your children on family trips to the library and have them get library cards and check out books. Further, have your children read to you. Have a "book club" time with your children. When they finish a chapter or a book, you can ask what the book was about, why a character behaved in a certain way, what they thought of the ending, and what a character might do next.

Children can learn about the magic of language, words, and stories in many ways in addition to reading. As noted earlier, you can tell your children stories about their family and culture. Children love to hear stories about their grandparents and ancestors and about relatives and family friends who live in distant places. Encourage your children to write notes to their grandparents and other relatives. Point out words to your children wherever you go with them: the grocery store, the pharmacy, or the gas station. Like with the Chicken Charlie story in this primer, be observant of opportunities all around you to engage in adventures with your children as you travel.

Learning Opportunities

Many community events can enhance learning. A number of community and religious organizations provide opportunities for children and their family members to engage in positive social and learning experiences. Libraries, museums, free concerts, and cultural fairs can be enjoyable for the entire family. Children will develop a love for cultural outlets, expanding their ability to understand and communicate with many people and cultures around them.

Other types of activities can encourage children to pursue STEM-specific interests. Some schools and communities hold "Maker Week," where they provide opportunities for children to create their own inventions or artistic and useful gadgets.

**

There are also bountiful summer and school year programs, many offered free of charge, that can ignite your students' genius and imagination! An example is a S.T.E.A.M CAMP (science, technology, engineering, the arts, and math) offered for the first time this summer at the Hamtramck Parks Conservancy. Executive Director and CEO, Dichondra R. Johnson, stated that the S.T.E.A.M Camp was sponsored, in partnership with the generous support of General Motors, PPG, and the GHD Foundation, to provide a summer learning experience in STEAM for students in the Hamtramck and Detroit Public Schools. Through this time of experiential fun and discovery, children, ages 9-14, played games as they learned about ecosystems, supply chain and product life cycles, and how to develop composting sites and make recycled materials. This program will greatly benefit children in the Hamtramck and Detroit communities and beyond. At this 3-day STEAM Camp the participants also learned about Circular Economy, creating treasures and 'nothing machines' from refuse and recycled items, the life cycle of a plastic bottle, the web of life and nature, and, of course, as they enjoyed a fun game of baseball in this, Hamtramck's historic baseball park! Projects like these set students' sights on unknown trajectories as they consider ways to utilize their own talents and resources in new and innovative ways, and seek the opportunity to share their learning experiences beyond school. Availing our children of these opportunities is wonderful for them and for parents as well!

SIGN UP FOR THE
S.T.E.A.M CAMP

The Hamtramck Parks Conservancy is partnering with GHD, General Motors, and Hamtramck Public Schools to run a free, 3-day STEAM Camp for students ages 9-14.

Hamtramck Parks Conservancy

2022 Hamtramck STEAM Camp!
July 12, 13 and 14 from 9:00 am - 11:30 am

Who: Hamtramck and Detroit Students, age 9-14!
What: The Hamtramck STEAM Camp!
Where: Veteran's Park, 8648 Joseph Campeau Ave Hamtramck MI 48212
When: Tuesday, July 12- Thursday, July 14. 9:00am-11:30am

The Hamtramck Parks Conservancy is partnering with GHD, General Motors and Hamtramck Public Schools to run a free, 3-day STEAM Camp for students ages 9-14. STEAM stands for Science, Technology, Engineering, Arts and Mathematics.

The Camp will be a fun, active learning space for students who are interested in Science, Arts and Sustainability around our parks. Lunch and t-shirt will be provided for every student.

Register here: https://hamtramckparks.com/em-event/steam-camp/

FREE 3-DAY CAMP
for Detroit and Hamtramck Public School Students
(Ages 9 to 14)
(Lunch and t-shirt provided)

Image Courtesy of Dichondra R. Johnson Executive Director and CEO of the Hamtramck Parks Conservancy, Hamtramck, MI, 2022

Photos Courtesy of Dichondra R. Johnson, Executive Director and CEO of the Hamtramck Parks Conservancy, Hamtramck, MI, 2022.

***

Colleges and universities have maker project opportunities for children who may be hoping to create STEM or entrepreneurial products in the future. These programs are often coupled with a mentorship opportunity. At the University of Michigan, for example, college students mentored third and fourth graders and encouraged them to make Legos into pianos or into moving structures, such as a Lego alligator. This was accomplished by connecting "LittleBits" electronic modules to Lego blocks, which was a new activity for children who customarily only worked with Legos. Parents can involve their children in these types of STEM skill-building programs, which will greatly benefit the child's knowledge of STEM culture and technology, build their skills and confidence, and perhaps offer them an opportunity to receive mentorship.

"Have a vision. Be demanding."
General Colin Powell

When observing your hobbies and skills, your children can learn vital information that will aid their STEM/STEAM interests and development. Make these times with your children into teachable moments. Each of the examples below contain many ways to teach children STEM-related skills and impart STEM knowledge to them as they help with home, agricultural, industrial, or environmental projects.

We can cherish, nurture and create deeper levels of learning, and even spice it up from time to time. Do something special and spontaneous with your children every chance that you can. Moreover, if you are currently looking to find new hobbies and fun things for yourself, make it a fun date that includes your children. The happier and more connected you are with them, the happier and more confident they will be in the world.

Home Routines and Practices

Studies show that successful students have parents who create and maintain positive family routines. Establishing a daily family routine with scheduled homework time, mealtime, bedtime, and setting aside time to examine your child's homework, is important. Review what he or she learned in school that day; is the homework correct, completed, and neat? Share reflections, problems, or concerns regarding your child's homework daily, asking for experiences he and or she might want to share regarding the homework. The "sharing time" can include you praising your child's successful grades and outcomes, and hearing about matters of concern. Pay special attention when your child relates anything about bullying or being bullied, drugs, alcohol, racial or gender discrimination, molestation, threats, or intimidation. Also, listen for signs of stress, tension, depression, or other concerns that require comfort, support, or professional help. A daily sharing time enables timely and supportive parenting, and it will be of great benefit to your children.

Interactions with Teachers, Counselors, and other Educators

Involve yourself in your child's school activities, help them with schoolwork, and seek learning enrichment and extra-curricular activities for them. You should be acquainted with your children's teachers in every grade. At the beginning of each semester, meet with your child's counselors and teachers and discover what courses they need at each grade level. Having a checklist to mark off as they complete each requirement is very helpful. In addition to the science and math courses they will need, including geometry, chemistry, computer technology, and advanced occupational courses, you can inquire about other courses that would be beneficial, such as history, science, the arts, geography, and foreign languages.

If your children exhibit gifted abilities, ask a counselor to involve them in the gifted and talented program. If the school does not offer a program, you can encourage the school to develop opportunities that would specifically benefit gifted and talented children. Being involved with your child's education is beneficial for everyone. When it comes to children succeeding in STEM, you should communicate engagingly with the teachers and administrators, and teach your children to do so as well. What about their teachers? Do they communicate openly with the parents and children? When children succeed in STEM studies, or if they have problems in the classroom, how do their teachers respond? Develop a comfort level in communicating with the school personnel in your children's academic and community settings, and they will feel supported.

Technology

Do your children watch more than seven hours of television a week? If so, it will be beneficial for you to choose selected educational television programs, and educational games for them to play. Also, choose programs and movies that you and your children can watch together, or play educational and mentally stimulating games together, and then review afterward what ideas, skills, stories, plots, and characters they might have enjoyed. A weekly screen time and TV viewing schedule can help you track the number of hours of television and electronic device gaming your child is consuming.

Ask your children's educators about Math Apps for their iPad, smart devices or computers that they can use at home to practice their skills. Some school libraries also lend smart devices for students' use. Some examples of applications include the following that are accessible on www.teachhub.com:

- Math Bingo (Math BINGO–selectable skill levels for grades K-5 and is similar to the traditional game of Bingo.

- Bugs and Numbers–Little Bit Studio, LLC, have wonderful graphics and depict bugs (insects) as actors in a story. Players earn the bugs by completing each of the skill games. Select the correct starting point in the game for your child, depending on their age and level of math development.

- Algebra-Your Teacher.com offers comprehensive math-specific apps for all K-12 levels. You will need to guide your children to the correct starting level.

- Math Board – Pala Software, Inc., allows a student to do math work in real time with a built-in interactive Live Learning Center.

- Numbler–Math Game, Brainingcamp, LLC–features a Scrabble-type of board interface that makes equations from numbers, rather than words from letters. Guide your student to the correct start point in the program.

- Another interesting program that has been shared in K-12 schools in East Lansing and Okemos, Michigan and in other K-12 school systems is Hour of Code, www.hourofcode.com, which introduces students to computer programming, a growing and evolving STEM profession.

- With the effects of a Covid-19 pandemic, war, race and domestic terrorism on the rise, we all seek answers. Many in STEM careers aid our ability to thrive through uncharted waters and unprecedented times. Math games, home-based math projects, and programs in elementary schools help students advance their math and science skills. The University of Michigan Youth Policy Lab visits elementary schools in Taylor (Wayne County), Michigan to help first graders play "Chocolate Chip Count," a fun game that teaches math skills. As noted by Deborah Hubbard, Research Assistant with UM Youth Policy Lab, game cards feature chocolate chips that the children count and then write their sums on a blank card. After they answer, the teacher tells them to feed the card to the monster face attached to a plastic jar. Another game called Stinky Socks; the children hang socks on a clothesline in order from 1 to 20. If they pull a stinky sock card, the cards are all pulled down and the game starts over again. These types of games teach basic math while the students are fun at play. What a great way to learn a lifelong skill. Liz Biddle, a school improvement coordinator for the Taylor school district suggested "High 5s," a math enrichment program developed by the University of Michigan, which is being implemented for kindergarteners and 1st graders at Randall and Myers Elementary Schools in Taylor. A study found the program increased kindergarteners' math performance by 15%. Moreover, as the name suggests, students are given plenty of positive reinforcement as they progress. These and other game programs aid future STEM students of color in closing the achievement gap.

Information Courtesy of Greta Guest, Michigan News, April 18, 2022, Vol 77, No. 28.

***

Physical, Mental, and Emotional Wellness

EVIDENCE-BASED RESEARCH HAS linked physical fitness to students' ability to excel in the classroom. Alexandra Sifferlin, in her article "Study: More Active Teens Get Higher Test Scores", Time, 10-22-13, noted that researchers from the United Kingdom compared students' activity levels with their academic performances in English, math, and science, and they concluded that the more moderate to vigorous physical activity the children had, the higher they scored on their academic tests. This finding was especially true for girls and their science scores. The effects also seemed to last, as the students grew older, and the link between more exercise and better science performance continued to be strong.

In other studies, researchers identified particular sports that help children the most, including, soccer, lacrosse, Tae Kwon Do, swimming, tap dance (or rhythmic, repetitive dance), ultimate Frisbee (getting the frisbee to the end zone without interception, much like football), tennis, hockey, wrestling and rock climbing (indoor or outdoor).

While exercising does not guarantee better grades or test performance, the evidence is clear that students perform progressively better, when they are physically active.

*"What mental health needs is more sunlight, more
candor, more unashamed conversation."*

Glenn Close

Children's mental and emotional health also has significant effects on their abilities to learn. Current brain research shows that children who are anxious about their learning, their homework, or other concerns have elevated levels of cortisol, the hormone associated with stress. A perpetual state of elevated cortisol levels can damage the body in numerous ways, and impair cognitive abilities, researchers have found.

Consider the following ways for providing a calming atmosphere for your children:

Make your home conducive to learning. Decorate your children's rooms with calming colors and effects, and provide a place for them to display their schoolwork proudly. Additionally, educators have found that "fidget toys" and exercise balls are good ways to help children express their energies before settling down to their homework.

Reduce loud and extraneous noise. From the time children arrive home from school, create transitions to help them wind down and prepare to study. First, encourage discussion and "lighter" activities, followed by family meals, study time, and relaxation, then bedtime. These transitions build in a sense of peace, continuity, and consistency in the household.

Help your children understand and appreciate their learning styles and increase their sense of ownership of the educational experience.

Helping Our Children When They Need a Friend

"Hold on to your dreams of a better life and stay committed to striving to realize it."

Earl G. Graves, Sr.,
Founder of Black Enterprise Magazine

The Day Carl Was Up to Bat

OUR MINISTER, DR. Stan Craig, at the Holmes Road Church of Christ in Lansing, Michigan, shared a moving story about a young man named Carl whose family came to see him play baseball. As the game began, Carl was not having a good day. When it came his time to bat, Carl swung, and everyone saw the ball fly into the catcher's mitt. Not only was Carl out, but the game was over and he was the cause of his team losing the game. In most situations, Carl would have gone home in a sense of defeat. While the winning team went crazy, their families swarming out onto the field to congratulate and celebrate their players, Carl's team walked quickly off the field to their cars and went home. Carl stood at the plate alone, sad and devastated. Then, he heard someone yell, "Come on Carl, pick up the bat, grandpa's pitching!" Bewildered, Carl looked around, slowly picked up the bat and swung at grandpa's first pitch, then the second, and on the third pitch, he hit the ball out to first base, where his mom missed the ball, saying it was due to the bright sun in her eyes. So, Carl ran to second base, where, amazingly, Uncle David, who seemed also blinded by the sun, missed the catch. 'Keep running, Carl', yelled another family member from the side as Carl headed for 3rd base, where the throw went at least two feet over the head of Cousin Joe, the third baseman. "Keep running, Carl!" they yelled on, and Carl raced for home base, running as hard as he had ever run. The ball was thrown with sure accuracy to the catcher blocking home plate, waiting to tag Carl out. However, just as Carl reached home plate the ball bounced in and out of Aunt Suzie's catcher's mitt, and Carl was home safe. Before Carl knew it, he found himself being swooped up and carried around on Uncle David's shoulders, while the rest of the family crowded around, cheering Carl's name!

A person observing this whole thing from across the field said, "Wow, I just witnessed a little boy become the recipient of grace". Carl's family had certainly come to his aid that day, and the warm and loving memory of family will certainly fuel his successes, and buffer any hard times, throughout his life. It is the grace of a good family, parents, communities, the 'village' that stands with us unconditionally, and this is a hopeful blessing for every STEM/STEAM student and professional.

Communicating About Learning

How do your children feel about school? Do they leave for school each day with a positive attitude, having had breakfast, a hug, and encouraging words? If they express or exhibit illness, fears, concerns, or apprehensions, do you assist them immediately and listen attentively to their woes? Children who are anxious or fearful about their learning, their abilities, or any other issues in their lives will have a difficult time performing well in their studies.

Whether your children are extroverts or introverts encourage them to communicate at their comfort level, and not be intimidated about expressing and communicating at school, at home, or in community settings. Students can perform well regardless to whether they are introverts or extroverts. Just as extroverts can excel, a study on the "power of introverts" concludes that a person with more introverted traits, when focused, can excel greatly in reading, writing, math, science, and reflecting, and their preference for small group interactions could lead to meaningful learning experiences. Introverts may be delighted to discover how well they can express themselves, and that the time they spend alone and deep into themselves is a part of their learning. In her book, "Quiet: The Power of Introverts in a World That Can't Stop Talking", 2012, author Susan Cain expressed that throughout history, introverts have played important roles as leaders, innovators, statesmen, and cultural icons, bringing special traits and benefits to our society by their thoughtful, reflective insights. She noted that some of the most transformative leaders in history, including Abraham Lincoln, Eleanor Roosevelt, Mahatma Gandhi, Bill Gates, and Rosa Parks were introverts. Charles Darwin, a historical STEM figure known as the Father of Evolution, was an introvert. He loved solitude, took long walks in the woods, and turned down dinner invitations. Dr. Seuss, another noted introvert, wrote alone and was afraid of meeting the children who read his books because he feared they would be disappointed in how quiet he was. Indeed, Albert Einstein, one of the most accomplished STEM scientists in human history, was characterized as an introvert who did his best thinking alone. "The monotony and solitude of a quiet life stimulates the creative mind," Einstein stated.

Ultimately, society needs balance. Introverts and extroverts help to even the experiences of human interactions. In the STEM/STEAM professions, there is a need for inventors, practitioners, and innovators. Introverts and extroverts could be those innovators waiting to emerge, and it is up to you to encourage your children to embrace their attributes and mold them into agents for change in the world of STEM. Encourage introverted children to read, to reflect their thoughts on paper, and to follow their interests. Invest time and resources in their interests, and if possible, devote time particularly to those hobbies that absorb and captivate them.

Cain also noted that people can be both introverts and extroverts, and these traits might exhibit themselves at various times. While children may be more outgoing in some settings than in others, it is important for you to accentuate their talents and the things that they do well. This type of parental support encourages more self-assurance, enabling children to project their abilities confidently in class, in the community, in multiple STEM/STEAM settings, and in their careers.

***

*"Hold fast to dreams, for if dreams die, life is
a broken winged bird that cannot fly."*

Langston Hughes

College Preparedness

To help your child prepare for college, you can ask teachers or counselors about dual enrollment at the local community college while your students are in high school. You can also ask about early college options, AP courses (advanced placement college readiness courses), and online learning opportunities in STEM/STEAM. When your children are ready for college, you can help them talk with school counselors and apply early to colleges and universities of their choice, and encourage them to seek scholarships and grants. During their college years in STEM, talk with college personnel and local businesses about potential internships for which they might qualify.

Investing in your child's college education and future career will help strengthen his or her economic future. According to the U.S. Department of Education NCES, 2015), students who earn bachelor's degrees earn an average of 66% more than those with only a high school diploma, and this outlook increases even more for STEM graduates.

Encourage your student to create a dashboard (progress report) on their computer or smart device to track their progress. Encourage your children to seek assistance from school counselors or advisors if they encounter any problems, need tutoring, or are feeling anxious or overwhelmed. Encouraging them to join support groups and social networks will engage them in fulfilling college experiences, improve their sociability, and promote their wellbeing.

Summary

Parents provide loving support, structure, and advice, in order to instill in their children an interest in science studies and careers from an early age. These interests can be sparked when your children are assisted by tutors, participating in special STEM/STEAM projects in the community, exploring the science of horticulture, optics, a high-tech crime unit, Legos, and wind tunnels, or being involved in summer science camps. A child's potential for academic curiosity is limitless. As they are growing and changing, their abilities will also grow and change. They learn to love scientific discovery. You can show them by example that small steps add up to large accomplishments. One day, with encouragement and commitment, and their steadfast work, their dream of being a STEM professional will be realized.

After having robust conversations with your children, you can make notes about their responses, thinking about how to refine and use these responses later to support their interactions with STEM educators, college faculty, advisors, and employers.

The next section of the primer contains a number of role-playing scenarios to help you and your students communicate about education and STEM.

***

"Never be limited by other people's limited imaginations."
Dr. Mae Jemison, first African American female astronaut

You may remember the saying, "A mind is a terrible thing to waste" from a television campaign run by the United Negro College Fund

(UNCF). In a landmark work, *"The Biology of Empowerment,"* 2005, Dr. Lee Pulos stated that genius is lying dormant in all people. Many brilliant scientists, technologists, and engineers are waiting to be discovered from among the millions of diverse young people who have the ability. Now is the time for your students to become a part of the growing number of students of color and women as they obtain college degrees in STEM/STEAM and pursue these professions.

Stay Inspired

You are a hero to your children. They look up to you with pride and admiration. You, in turn, must be your children's champion and greatest ally. STEM/STEAM parents have a precious opportunity in your hands. Magnify and focus on the positive aspects of parenthood. Parents may lament that their educational background, race, ethnicity, or socioeconomic status is unlike communities of wealth, but their children only care about the person they want to emulate and become. Children say, "When I grow up, I want to be just like my dad, or I want to be like my mom." Children admire the people who love them and care for them, unconditionally, every day.

As you develop an appreciation and fascination for STEM/Arts education and all that it represents, your children will also catch your vision and develop a fascination for it, particularly if they have an inherent interest and talent in STEM/STEAM careers. The more curious children become about STEM topics, the easier it will be for them to learn information about them. Research has shown that brains develop new pathways when curiosity becomes piqued about a subject. On the other hand, if children express that they do not want to pursue STEM careers, and would rather be a painter, chef, or many other talented roles, parents can still encourage them to have an understanding of science and technology as they complete academic studies in a field of their choice. Retention and completion are important, and should be among your children's goals, even as their ultimate career interests evolve.

Going forward, parents of diverse students must take a proactive role in guiding, coaching, and mentoring children in all of the learning environments that they experience, from pre-K to college. Be their advocate, be their listening ear and their protector, and most of all, be their "fan" and encourager. If you believe in your children, they will believe in themselves.

The following are examples of potential quotes that students might state as they describe their experiences:

- "My dad and mom are always there at my Saturday Morning Physics events. I always get a jolt of confidence as soon as I see them walk through that door! Even though I am the only person of color in the room, I feel like I am somebody because my parents are there, and they believe in me, unconditionally."

- "I feel down some days, but my mom always says to me to pray about it and believe that all will go well, and, you know, often it's not long before I start feeling better!"

- "Grandma and Grandpa never went to college, and my dad and mom dropped out of high school to work and help support their family. However, every time I am invited to the Math Quest, where I have gotten first place for the last two events, they are there to support me, and that makes me feel so good. I have a full ride scholarship to the Massachusetts Institute of Technology (MIT). I hope my grandparents and parents will live to see my graduation from MIT. I am going to love seeing the pride on their faces and the tears in my mother's eyes!"

There may be times as a parent of students of color, female or diverse students, that you wonder if your efforts are effective. You may wonder if you are ultimately making a difference in your children's progress. Rest assured, as a proactive STEM parent, you are effective, and you are in charge of shepherding your children through the early development of their STEM/STEAM journeys. All efforts that you make in this quest will be invaluable to them. In the next chapters, we will explore specific exemplary programs that can aid children's STEM/STEAM education and career development.

Role-Play Exercises to Sharpen STEM Discussions and Readiness

THESE ROLE-PLAY SCENARIOS, for diverse students, will provide opportunities for practice ahead of time so that the conversations you and your children have with educators and mentors will be optimally engaging, personable, productive, and meaningful. In these exercises, you can assume the role of a STEM instructor, school administrator, or school counselor who will ask these general questions of your children for purposes of eliciting responses and gaging your children's interest. As well as, role-play exercises are included for older children to assume the role of school personnel who might ask questions of the parents. You and your children can also use these as personal preparatory questions as you prepare to meet and converse with educators and STEM professionals.

Role Play Questions to Ask Students in Grades K-12

- What is your name? Tell me about your life to date. Is there anything you would like to share about your family, school, home, and friends?

- Who, or what, inspires you in math and science?

- What career would you like in STEM/STEAM? Would you like to see some of the many STEM and Arts fields you might like to pursue?

- Tell me about your parents and grandparents (and other senior family members) their education, careers, hobbies, and something about them that is especially wonderful?

- Do you have siblings? If so, describe their personal interests, hobbies, school, and careers.

- Who or what inspires you to do well in school? Do you get good grades in school?

- Who is your favorite teacher? What do you enjoy about your teacher?

- *Who do you think will help you go to college?*

- Are your parents important to your schooling and ultimate career success?

Role Play Questions to Ask College Students

- Have you experienced barriers, challenges, or setbacks in your quest to succeed in your STEM college courses, or in your pre-career endeavors? Describe the challenges and what ways you sought to overcome them.

- What do your parents or mentors do to aid your progress, or to assist you in handling the challenges?

- What recommendations would you provide parents of diverse students that might aid them in guiding their children to successful outcomes in formal STEM training and STEM career attainment?

- What advice would you give aspiring young people of color (African American, Latino, and Native American), first-generation students, and women who are seeking STEM educations and professions?

- How would you rate your critical thinking skills, and how have they aided your STEM career path?

- What were your current and previous grades in your math and science courses?

- Tell me about the STEM career you are pursuing and about the excitement, you are experiencing and anticipating in achieving your dream.

Role-Play Questions to Ask Parents

- What is your name? Tell me about your life to date. Do you have anything you would like to share about your background, ethnicity, culture, upbringing, family, school life, job, career, home, children, or other topics?

- What were your career aspirations growing up? Did you accomplish your goals? What are current dreams and aspirations you are pursuing that you would like to share?

- Who, or what, inspires you?

- Have you ever had an interest in STEM areas, courses, related hobbies, or as a profession?

- If you were to pursue a STEM career, what would you pursue (can be answered either from a list of STEM/STEAM professions, or by general description)?

- Tell me about anyone among your family and friends who works in STEM education, or STEM-related professions?

- Have you experienced barriers, challenges, or setbacks in your quest to succeed in your career or in life? Describe the challenges, and what ways you have sought to overcome them.

- Tell me about your children, their ages, and grades in school. What aspirations do you have for your children's education and their future? Do you encourage them to work hard in school?

- Do/did you read to your children? If so, did you begin reading to them from birth? Do/did you place them in an early child development or preschool program?

- Do you see yourself as a mentor for your children?

- Do you inquire about opportunities that will aid your children's progress? Do you assist them in handling any barriers or challenges they encounter either at school or in the community? Tell me about some of the avenues you have pursued.

- Do you accompany your children to school activities and attend parent-teacher conferences?

- Do you encourage and accompany them during their outlets and hobbies (math and science clubs, music, soccer, boy or girl scouts, martial arts training) that encourage their social and academic development?

- Do you seek programs at school or in the community to aid their academic knowledge and growth?

- Do you request academic tutors or study pals/groups for your children at school or in the community to aid their understanding of subjects with which they are seeking assistance?

- If you could request assistance in mentoring your children to improve your skills at furthering their academic understanding, growth, and success, what types of information, training, resources, or academic provisions would you find most useful?

- What recommendations would you provide other parents of diverse students that might aid them in guiding their children to successful outcomes in formal STEM training and STEM career attainment?

• What advice would you provide aspiring young people; including your own, in diverse populations (African American, Latino, and Native American), first-generation students, and women who are seeking STEM education, and ultimately who hope to succeed in STEM professions?

You and your students can continue to reflect on these questions and add others that stimulate an appreciation for your family's goals and dreams. This reflection can improve confidence as you advocate for your children in their STEM/STEAM pursuits, and can improve their confidence in communicating their STEM interests.

***

STEAM Resource Guide

"Success is a journey, not a destination."

Arthur Ashe

Introduction

MANY EXEMPLARY STEM programs and opportunities are available for your children to explore. Academic and cultural resources exist in school districts, colleges, and communities nationwide. Through your student's instructors, counselors, gifted and talented programs, community resources, local community colleges or universities, and parent groups you can learn about empowering programs, and many are free of charge or low cost.

Whether or not your children are receiving the all of the STEM training you are seeking in their current schools, these programs provide more options of the many academically fulfilling opportunities available for your children.

Compendium of Academic Programs and Resources for STEM

These programs are in various geographic locations in the U.S., and serve as examples of opportunities you may find available in your locale. Reach out to your children's teachers and school counselors to become knowledgeable about special STEM/STEAM learning opportunities and programs your children will enjoy.

Examples of Programs that Enhance STEM/STEAM Development

PK-8 Excellence. Harrison Park Academy, Grand Rapids, Michigan. Harrison Park Academy has historically been an excellent school for young scholars of all nationalities.

In May 2022, Davenport University and the Grand Rapids Public Schools signed an agreement to address the teacher shortage, and provide 75 new scholarships for urban STEM education degrees for current students within the district.

Harrison Park School is an example of the kinds of excellent teaching environments that exist, that are cost-effective for students of all socioeconomic groups, and that engage the parents to participate, regardless of their backgrounds.

Contact information:

Harrison Park Academy, Heather Thompson, PhD, Principal
1440 Davis Ave. NW, Grand Rapids, MI 49504
Phone: 616-819-2565
Fax: 616-819-2567
Website: https://grps.org/schools/elementary/harrison-park/

***

K-8 STEM for Students at Boren in Seattle, Washington

K-8 STEM at Boren provides STEM courses for students in kindergarten through 8th grade. There is academic support for students at all levels and special programs for advanced students. The K-8 STEM curriculum is project-based and is designed to help the students develop critical thinking skills as they prepare for the future. Washington State is increasingly becoming a high-tech economy. Washington STEM can be contacted at: https://washingtonstem.org to learn more about their career STEM education programs for historically underrepresented students.

Contact Information:

K-5 STEM at Boren
5950 Delridge Way, Seattle, WA 98106
(206) 252-8450
Website: https://borenstemk8.seattleschools.org/

***

This Boys and Girls Young Entrepreneurs Academy

Special summer and advanced programs for middle to high school students of color are an excellent way to engage your children in experiential learning. This Boys and Girls Young Entrepreneurs Academy, for ages 10 through 17, is sponsored by the T-Rose Foundation and a number of corporate sponsors. It is offered at the Marygrove Conservancy in Detroit MI. The students learn how to create a business plan, business registration, professional headshots, and a business pitch video.

This event offers free admission and dorm room accommodations.

Image Courtesy of Theresa Randleman, Founder and CEO of
the T-Rose Foundation, Michigan 2022 www.t-rose.com

***

Focus: HOPE, Detroit, Michigan. Focus: HOPE is a community-based
resource center founded in 1968 by Father William Cunningham
(1930–1997) and Eleanor Josaitis (1931-2011), dedicated to providing practical solutions to problems of hunger, economic disparity,
inadequate education, and racial inequity. As shared by the excellent
leaders at Focus: HOPE, it's programs provide early learning, STEM
and industrial career programs for underserved people in Detroit,

from pre-school-aged children to seniors. For college-aged students, it provides technical skills training that lead to engineering degrees at Lawrence Technological University, Wayne State University, the University of Detroit Mercy, the University of Michigan-Dearborn, or the University of Michigan-Ann Arbor.

The Center for Advanced Technologies (CAT) at Focus: HOPE provides a unique learning experience for students interested in becoming engineers. Students begin by enrolling in a STEM Bridge course to advance their STEM skills while earning college credit. Upon successful completion of the STEM program, they enroll at Lawrence Technological University to earn an associate's degree in manufacturing engineering technology. The students receive "wrap-around services" while pursuing engineering education at the partner universities. They have the added benefit of exposure to the professional "world of work" through internships with area employers.

Along with a High School College Bound program, Focus: HOPE provides an Information Technologies Training Center. IT operations train students for industry certification, and offer a machinist training program for those wishing to become a machine operator.

Contact Information:

Focus: HOPE
Early Education and Center for Advanced Technologies
(CAT) Programs
1400 Oakman Boulevard
Detroit, Michigan 48238-2848
(313) 494-5500
Website: https://focushope.edu/

***

GEAR UP. The University of Michigan's GEAR UP (Gaining Early Awareness and Readiness for Undergraduate Programs) provides an opportunity for underrepresented and first-generation students to discover their potential, gaining knowledge and skills they need to complete high school and prepare themselves adequately for college entry and success in STEM.

GEAR UP services begin in middle school, guiding students and families through the early years of college preparation, including the academic, social, emotional, and financial aspects of preparation. U-M Ann Arbor GEAR UP seeks to empower students and ensure their academic success as they proceed to college.

Images Courtesy of UM Center for Educational
Outreach (CEO), Ann Arbor, MI 2022

103

Contact Information:

GEAR UP
Center for Educational Outreach (CEO)
1214 South University Avenue, 2nd Floor
Galleria Building, Room 248
Ann Arbor, Michigan 48104-1316
Phone: (734) 647-1402
Website: https://ceo.umich.edu/gearup

***

DAPCEP (Detroit Area Pre-College Engineering Program) provides STEM educational programs for children and teens. It connects metropolitan-Detroit youth with educational opportunities in STEM. On Saturdays during the fall and spring, and in camp format over the summer, DAPCEP places youth on university campuses to learn various subjects as, renewable energy, engineering, computer programming, and the mathematics of music. All underrepresented students in pre-kindergarten through 12th grade are welcome to participate. The Michigan schools and universities with which DAPCEP partners include the following:

- Oakland University
- Lawrence Technological University
- Michigan State University
- Michigan Technological University
- University of Detroit Mercy
- University of Michigan, Ann Arbor

- University of Michigan, Dearborn
- Wayne State University
- Detroit Public Schools
- Southfield Public Schools

DAPCEP has been preparing students in STEM since 1976. Its alumni have gone on to become engineers, doctors, entrepreneurs, computer scientists and more.

Contact Information:

DAPCEP
2111 Woodward Ave., Suite 506-1
Detroit, MI 48201
Phone: 313-831-3050
Fax: 313-831-5633
Website: https://www.dapcep.org/

***

Whether you explore code.org, codeacademy.com or so many others, there are a plethora of opportunities to help your children learn coding, as early as pre-school or kindergarten! Because the concepts of coding can easily be presented in fun games, block-based puzzles and sequencing, pre-readers can learn the basics right along with older children and parents. For ages, 5-7, drag and drop games can help with basic coding concepts.

In coding, computer programmers use various programming languages to "tell" computers, mobile devices and software programs how to run. Websites can be developed, for instance, using HTML and CSS, and games developers might use Python or Java.

Using coding guides, you and your children can immerse in the exciting learning models available on coding, computer programming, even weather and nature tracking. The benefits of this time of exploration will be greatly realized by you and your family, particularly as your children enter STEM careers. Start your kids coding today!

*****

City Year Detroit—Student Mentorship

City Year Detroit is an empowerment organization dedicated to improving student success in the Detroit Public Schools, in partnership with government and corporate sponsors. City Year Detroit is recruiting young adults, 18-25 years of age, to serve as mentors/tutors in Detroit Public Schools. This opportunity includes a biweekly stipend, education award, access to scholarships, professional development, and direct access to job opportunities. This is an example of the many programs that are connecting to aid underrepresented youth in educational and STEM opportunities.

Contact form https://pages.cityyear.org/Allison-Knox.html

Image Courtesy of City Year Detroit, Detroit, MI, 2022

***

Discovery Education provides school districts across the nation with digital learning media in standards-based content. Their products, which are available on the Discovery Channel, have been proven to influence students positive achievements. It is important and helpful to expose children to media that engage them in STEM. The

Discovery Channel and Discovery Education have stated that they are on a mission to propel today's students, schools, careers, and citizenship. Parents can ask their children's teachers about the potential of incorporating Discovery Education digital learning media, or similar digital learning tools, into their curricula. The Discovery Education network online lists the types of programs parents can access at home for their children.

Website: www.discoveryeducationglobal.com
Phone: 1-800-323-9084
Fax: 1-855-495-6542
Email: Education_Info@DiscoveryEd.com

***

Michigan Virtual Charter Academy

This tuition-free online public school offers individualized learning, an interactive curriculum, educational field trips, social events and clubs, and high school STEM and AP courses for college credit.

Contact information:

Michigan Virtual Charter Academy
Phone: 855.514.2349

***

Community College Innovation Challenge Awards for STEM Students

In the fall of 2014, the National Science Foundation launched the Community College Innovation Challenge (CCIC). This project involves teams of community college students and mentors proposing innovative science, technology, engineering, and mathematics STEM-based solutions for real-world problems that they identify within a variety of areas, including:

- Big Data
- Sustainability (to include water, food, energy, and the environment)
- Infrastructure Security
- Broadening participation in STEM

Contact Information:

American Association of Community Colleges
One Dupont Circle
Washington, DC 20036
E-mail: ccic@aacc.niche.edu
Website: www.aaccinnovationchallenge.com
Phone: 202-728-0200 x217

***

First2 Network

The First2 Network is a grassroots program in West Virginia that supports first-generation college students in STEM. It aims to improve the college enrollment rate of first-generation students, success of undergraduate STEM students, particularly those in rural communities. Initiatives that support their mission include: improving the STEM college experience, connecting successful programs that improve persistence, fostering new partnerships, conducting new research, and providing overall wrap-around services to aid first-generation students. The National Science Foundation funds First2 Network.

Contact Information:

https://first2network.org

***

Lake Superior State University Robotics Undergraduate Program

In addition to four-year and graduate degree programs in STEM education, additional options provide training that meets the current trends and demands of innovation. Companies involved in robotics and automation regularly seek LSSU graduates. In addition to robotics, LSSU offers associate's degrees in General Engineering, General Engineering Technology, and Manufacturing Engineering Technology. LSSU also offers Bachelor of Science degree programs in all areas of engineering and industrial technology.

Contact Information:

Lake Superior State University
School of Engineering & Technology
650 W. Easterday Ave.
Sault Sainte Marie, MI 49783
Phone: (906) 635-2231
E-mail: admissions@lssu.edu
Website: https://www.lssu.edu/college-innovation-
solutions/school-engineering-technology/
engineering-degree-programs/robotics/

***

Youngstown State University (YSU) Summer Bridge College-based Programs

Youngstown State University's Center for Student Progress Summer Bridge is a free, one-week residential program for first time, tradition-al-aged, multicultural freshmen entering Youngstown State University, in Ohio. It is designed to provide students the opportunity to become more familiar with the academic and social experiences most often encountered by first-year students. During the summer, students in this program reside in YSU residence halls; take introductory Writing, Study Skills, and Computer mini courses; participate in presentations from Financial Aid, Student Life, Campus Recreation, and Student Employment; and participate in social activities within the city of Youngstown and surrounding communities.

What makes this program unique is that it transitions to Bridge and Beyond during fall semester. During this aspect of the program, participants have block scheduling of classes, live in a learning community in the residence hall, have a peer mentor, and work weekly with a professional staff mentor who serves as an academic coach. Students who have participated in this program have greater academic success and graduation rates than their peers. Youngstown State University is an urban, open-admission, 4-year public institution located in Northeast Ohio.

Contact Information:

Center for Student Progress
Youngstown State University
One University Plaza
Youngstown, Ohio 44555
Phone: 330.941.3000
Website: https://ysu.edu/diversity-and-inclusion/summer-bridge-program

***

Kent Intermediate School District, Grand Rapids, Michigan

Kent Intermediate School District connects students to mentors, student organizations, and work-based learning opportunities. They have an aviation program and career technical programs.

Contact Information:

KENT ISD/Kent Career Tech Center
2930 Knapp St. NE
Grand Rapids, MI 49525
616-364-1333
http://www.thetechcenter.org/

***

Lansing Community College Aviation Program

The state-of-the-art LCC Aviation Maintenance Technology Center is located at the Jewett Airfield in Mason, Michigan. This aviation program is considered be the best in the Midwest.

All Lansing Community College students learn on industry-current equipment while working on large and small planes, turbine and jet engines, traditional aircraft construction and advanced composite materials. The facility is equipped with classrooms, a computer lab, and separate bay workstations for 1-on-1 and group instruction. Graduates in the various aviation specialties graduate in two years, and are sought after by Delta Airlines and many other companies. Delta partnered with the Lansing Community College several years ago as it sought sites around the nation to train specialized aircraft maintenance personnel. Through hands-on, real world training in Basic Aviation, Airframe Maintenance, and Powerplant Maintenance, students are prepared to earn their FAA license after successfully passing the FAA exam. Many students receive excellent employment offers before they complete the program. Other graduates choose to transfer to a four-year institution to continue for a higher degree, after completing this program.

Photo Courtesy of the Lansing Community College
Aviation Maintenance Technology Center, 2022

Contact information:

LCC AVIATION MAINTENANCE TECHNOLOGY
661 Aviation Drive
Mason, MI 48854
Phone: 517-483-1406
Email: aircrafttech@lcc.edu
Website: https://www.lcc.edu/academics/areas-of-study/ computers-engineering-technology/aviation/index.html
Facebook: LCCAviationTech

***

Summary

These STEM learning models shared are an example of the many programs that you and your children can explore nationwide

The most important point to remember when reflecting on these models is that you are in charge of your children's educational experience. In preparing them for STEM/STEAM education and careers, you are empowered to involve them in the best educational opportunities available.

STEM skills are in demand, as shown below in Figure 1.

Figure 1: Employment Comparisons for STEM and non-STEM Occupations
Source: www.changetheequation.org/stemdemand

STEM Careers

Following is a representative listing of STEM/STEAM Careers, compiled from a variety of sources to provide you a representative sample of the many STEM and STEAM careers to which your students might aspire. Definitions are provided where helpful:

- Accountant

- Actuary: Analyze the monetary costs of risk and uncertainty using mathematics, statistics, and financial theory.

- Aerospace Engineers: Design, construct, and test aircraft.

- Agricultural Engineers: Solve problems that are related to the way farms work.

- Airframe Maintenance Technician: Service, maintain and overhaul aircraft components and systems, including the airframe, piston engines, turbine engines, electrical systems, hydraulic systems, propellers, instrumentation, and warning and environmental systems.

- Airport Planner: Design conceptual airfield and lead landside planning, Business Park Development, rezoning, platting and development plan review, coordination of planning within FAA guidelines.

- Anthropologists and Archeologists: Study the behavior of human beings in different parts of the world and different periods in time.

- Applied Mathematician

- Architect: Plan and design buildings, aid in the construction planning and implementation and oversee the craftsmanship.

- Architectural and Engineering Managers: Coordinate and manage the work of architects and engineers.

- Astronomers: Observe and study stars, planets, and other astronomical phenomena.

- Atmospheric and Space Scientists: Investigate weather-related phenomena to prepare weather reports and forecasts for the public.

- Avionics Technicians: Install, inspect, and test electronic equipment in aircraft.

- Biochemical Engineers: Develop products using knowledge of biology, chemistry, or engineering.

- Biochemist and Biophysicist: Study the chemical composition or physical principles of living cells or organisms.

- Bioinformatics Scientists: Conduct research using bioinformatics theory.

- Biological Technicians: Assist biological and medical scientists in laboratories.

- Biologist

- Biomedical Engineer: Solve problems related to biological and health systems.

- Cartographers and Photogrammetrists: Collect, analyze, and interpret geographic information to study and prepare maps.

- Certified Nursing Assistant: Works directly with patients in a variety of care services under the supervision of a Registered Nurse, in hospitals, assisted living facilities, nursing homes, or home care.

- Chemical Engineers: Design processes for manufacturing chemicals and related materials.

- Chemical Technicians: Work in labs and assist with analyzing chemicals and other substances.

- Chemist: Conduct research on chemicals using experiments or observation.

- Civil Engineering Technicians: Use principles of civil engineering to plan and design construction projects.

- Civil Engineer: Perform engineering duties in the planning and designing phase of construction projects.

- College Anthropology and Archeology Teachers: Teach college-level courses in anthropology or archeology.

- College Atmospheric and Space Science Teachers: Teach college-level courses in the physical sciences.

- College Biological Science Teachers: Teach biology courses at the college level.

- College Chemistry Teachers: Teach college-level courses in chemistry.

- College Engineering Teachers: Teach engineering courses at the college level.

- College Environmental Science Teachers: Teach college-level courses in environmental science.

- College Geography Teachers: Teach college-level courses in geography.

- College Physics Teachers: Teach college-level physics courses.

- Computer Programmer

- Computer Scientist

- Construction Management

- Data Scientist

- Database Administrator

- Database Engineer

- Dentist

- Economist

- Electrical Engineering Technologists: Assist electrical engineers in a variety of activities.

- Electrical Engineers: Research, design, and test electrical equipment and systems.

- Electrical and Electronics Drafters: Prepare diagrams that are used to create, install, or repair electrical equipment.

- Electromechanical Engineering Technologists: Assist electromechanical engineers in a variety of activities.

- Electronics Engineering Technologists: Assist electronics engineers in a variety of activities.

- Electronics Engineers: Design or test electronic components for commercial, military, or scientific use.

- Emergency Medical Technician: Provide Ambulance and emergency medical services to aid emergency needs of patients and quickly transport them to a hospital or available medical facilities.

- Energy Engineers: Design programs or systems to make buildings more energy efficient.

- Environmental Engineering Technicians: Apply principles of environmental engineering to operate and test equipment that is used to help the environment.

- Environmental Engineer: Design ways to prevent and control pollution.

- Environmental Science Technicians: Perform laboratory and field tests to monitor the environment and find sources of pollution.

- Environmental Scientists: Research ways to remove hazards that affect people's health or the environment.

- Epidemiologists: Investigate the causes of health problems in communities or societies.

- Financial Analyst

- Forensic Science Technicians: Assist in investigating crimes by collecting and analyzing evidence.

- Fuel Cell Technicians: Work with fuel cell systems in a variety of settings.

- Geneticists: Research and study how traits are inherited from one generation to the next.

- Geographers: Study the earth and its land, features, and inhabitants.

- Geographic Information Systems Technicians: Assist scientists or others who use geographical information systems (GIS) databases.

- Geologist

- Geoscientists: Study physical aspects of the earth, such as rocks, soils, and other materials.

- Geospatial Information Scientists and Technologists: Research or develop geospatial technologies.

- Health and Safety Engineers: Promote worksite or product safety by using knowledge of industrial processes.

- Historians: Research and understand the past by studying a variety of historical documents and sources.

- Hydrologists: Study water that is underground or at the surface of the earth.

- Industrial Engineering Technicians: Help industrial engineers to design processes to make better use of resources at work sites.

- Industrial Engineering Technologists: Assist industrial engineers in a variety of activities.

- Industrial Engineers: Create systems for managing production processes.

- Manage Information System Analyst

- Manufacturing Engineering Technologists: Work to ensure effective manufacturing processes.

- Manufacturing Engineers: Design ways to improve manufacturing processes.

- Marine Engineers and Naval Architects: Evaluate materials and develop machinery to build ships and similar equipment.

- Materials Engineers: Develop ways to create materials for certain products.

- Materials Scientists: Research and study the properties of different materials, such as metals, rubber, ceramics, polymers, and glass.

- Mathematical Technicians: Use numbers to help solve problems and conduct research.

- Mathematician: Solve problems or conduct research using mathematical methods.

- Mechanical Drafters: Prepare diagrams of machinery and mechanical devices.

- Mechanical Engineering Technologists: Assist mechanical engineers with various tasks and activities.

- Mechanical Engineer: Use engineering principles to design tools, engines, and other mechanical equipment.

- Mechatronics Engineers: Research or test automated systems or smart devices

- Medical Scientists: Conduct research to understand human health and disease better.

- Microbiologists: Study the growth, structure, and development of very small organisms.

- Microsystems Engineers: Research, develop, or test microelectromechanical systems (MEMS) devices.

- Molecular and Cellular Biologists: Research and study cell functioning.

- Nanosystems Engineers: Apply principles of nanotechnology to develop specialized materials or devices.

- Nanotechnology Engineering Technicians: Operate equipment to produce or test materials at the molecular level.

- Nanotechnology Engineering Technologists: Create processes to produce materials or devices, working on an atomic or molecular scale.

- Natural Sciences Managers: Coordinate activities in a variety of scientific fields, such as life sciences, physical sciences, mathematics, or statistics.

- Non-Destructive Testing Specialists: Test the safety of various types of structures using x-ray, ultrasound, or fiber optic equipment.

- Nuclear Engineers: Conduct research on nuclear energy and nuclear waste disposal.

- Nuclear Technicians: Assist physicists, engineers, or other scientists with nuclear testing or research.

- Nurse Practitioner: Perform routine and advanced patient examinations, prescribe medications, order diagnoses, evaluate medical and test results, supervise nurses and other staff members, create individual treatment plans, record and insure accurate patient medical records, collaborate with other healthcare professionals to share knowledge, create treatments and diagnose patients, can work in a variety of medical settings.

- Operating Engineers and Other Construction Equipment Operators: Operate one or more types of power construction equipment.

- Paramedic: Registered healthcare professional who works autonomously across a range of health and care settings. May specialize in medical practice, as well as in medical education, leadership, or research.

- Park Naturalists: Plan and conduct programs to educate the public about national, state, or local parks.

- Petroleum Engineer: Create methods to improve oil and gas extraction and production.

- Pharmacist: Control, formulate, preserve and dispense medications at a pharmacy, and provide advice on how medications are to be used to achieve maximum benefit, minimum side effects, and to avoid drug interactions.

- Pharmacy Technician: Health care provider who performs pharmacy services under the supervision of a licensed pharmacists.

- Photonics Engineers: Design technologies involving light, such as laser technology.

- Physician (and allied Health and Medical professions)

- Physician's Assistant: May diagnose illnesses, develop and manage patient treatment plans, prescribe medications, and may serve as a principal health care provider. They have advanced medical training beyond nursing, but not a medical degree, and in many states P.A.s have a direct agreement with a physician. They typically cost health care organizations half as much as physicians, but perform many of the same services for patients.

- Physicist: Conduct research on the physical world by using observations or experiments

- Powerplant Maintenance Technician: Installation, maintenance and repair of mechanical, electrical, electronic, pneumatic equipment used in the generation of electrical power; may work in aviation.

- Psychologist

- Registered Behavior Technician or Supervisor: Assist in delivering behavior analysis services. The RBT Technician works under the supervision of an RBT Supervisor.

- Registered Nurse: Assess and observe patients, identify their critical needs, record patient medical history and symptoms detail, administer medications, care for wounds, draw blood, educate the patient and family on treatment and care plans, supervise licensed practical and vocational nurses, and additional duties in various settings.

- Quality Control Analysts: Conduct tests to study the quality of raw materials or finished products.

- Recordkeeping Weighers and Checkers: Weigh, measure, and check materials, supplies, and equipment in order to keep records.

- Remote Sensing Scientists and Technologists: Study a variety of topics using remote sensing techniques.

- Remote Sensing Technicians: Assist scientists using remote sensing techniques.

- Robotics Engineers: Research, design, and test robotic systems.

- Sales Engineers: Sell complex and scientific products or services.

- Social Science Research Assistants: Assist social scientists in laboratory, survey, and other social science research.

- Software Engineer

- Solar Energy Systems Engineers: Evaluate and analyze sites for solar energy systems, such as solar roof panels or water heaters.

- Statistical Assistants: Prepare statistical data for reports.

- Statistician

- Surveying and Mapping Technicians: Help surveyors keep boundary lines clear in construction projects.

- Systems Analyst

- Technical Writer

- Wastewater Treatment Plant Operator: Performing chemical treatment of oil, metal working lubricants, grease, coolants, acids, alkalis, and other waste products to meet pretreatment discharge requirements, performing all aspects of wastewater collecting, sampling, monitoring, testing required to maintain compliance with federal, state, and local regulations.

- Web Developer

STEM Arts (STEAM) Careers

Here is a representative sampling of STEM Arts Careers that combine STEM and creative skills:

- Robotics Engineering

- Landscape Architect

- Science Documentary Producer

- Software Developer

- Computer Hardware Engineer

- Biomedical Animator

- Civil Engineer

Soft Skills that aid STEAM Student Development and Success

Students, in your STEM/STEAM career, you will develop and implement new ideas, solve problems, and conduct research. You may work in a variety of settings including a laboratory, research facility, a classroom, office, or in the field. In addition to specific technical skills, STEM/STEAM jobs call for the following Soft Skills, including: analysis, good written and oral communication, attention to detail, problem solving, critical thinking, creativity, leadership, organization, and advanced science, technology, engineering, the arts or math knowledge. Developing these Soft Skills will greatly aid your educational and career success.

Epilogue

*"Those who say it can't be done are usually
interrupted by others doing it."*

James Baldwin

AS I HAVE shared, we as parents do not have to be a PhD or world-re-nowned STEM or STEAM scholar to raise one. We just need to have a willingness to see our children through their educational and career development journey. Coach, guide, love, and endure with them. I guarantee you, it will all be worth it - establishing a pathway of success and achievement for the brilliant, talented STEM and STEAM professionals that they will become!

Complete - The Man We Know

Time well spent, Energy given, Countless memories
A knowledgeable man accepts leadership responsibility
Build within and allow motivation for others
The man we know believes in Growth

Reaching out whether near or far
Maintaining contact strengthens relationships
Keep it real, real love, real smiles, real caring that you can see
The man we know believes in Family

Going out, being active, telling jokes
Celebrating, enjoying sports, laughing out loud
Living life, being around the ones you love
The man we know believes in Fun

Security lies in faith
Following a straight path, one has never lost his way
Kept by a higher power, the foundation for his values
The man we know believes in God

Because of the man we know
We believe in all these things

Because of the man we know
We have grown to be people of limitless potential

Because of the man we know
We are complete

The man we know has many roles
Whatever he is to you, we can all agree
Because of him we are complete

Melvin Johnson - The man we know
He is complete. We are complete.

Happy Fathers Day
2006

This poem was written by Poet and STEAM professional, Jonell Henry Johnson, in tribute to his father, Melvin Johnson, 2006

Children are a gift from the Lord; they are a reward from Him.
Psalm 127:3

***

For Parents and Families—A note of wisdom:

- Your job is the dream of the unemployed.
- Your house is the dream of the homeless.
- Your smile is the dream of the depressed.
- Your health is the dream of those who are sick.
- Beautiful things happen when you distance yourself from negativity.
- Your children are future scientists, technologists, engineers, artists and mathematicians.
- Find something to be grateful for.
- Enjoy the Blessings in all that you do.

***

For parents, family members, education settings and community partners seeking to affiliate with our Parent Empowerment Network (PEN), or procure assistance in guiding your students into STEM/ STEAM careers, we are available to provide consultation, advocacy, and mentoring.

Contact: Dr. Veronica A. Wilkerson Johnson, Executive Director
VWJ STEAM VENTURES, LLC
E-mail: info@vwjsteamventures.com

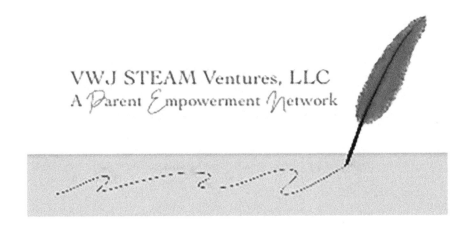

VWJ STEAM Ventures, LLC is an educational resource company dedicated to providing information, mentoring, consultation and advocacy for diverse, first-generation, and female students seeking academic training and careers in a multitude of disciplines in the Sciences, Technology, Engineering, the Arts, and Math (STEAM). Together we will foster excellence in your students as they explore their talents and achieve their dreams ~ the Sparkle in Their Eyes!

***

Milton Keynes UK
Ingram Content Group UK Ltd.
UKHW050248280224
438578UK00004B/21